# Agricultural Sustainability

# Agricultural Sustainability

## Strategies for Assessment

Gary W. vanLoon
S.G. Patil
L.B. Hugar

**SAGE Publications**
New Delhi / Thousand Oaks / London

Copyright © Gary W. vanLoon, 2005

All rights reserved. No part of this book may be reproduced or utilised in any form or by any means, electronic or mechanical, including photocopying, recording or by any information storage or retrieval system, without permission in writing from the publisher.

First published in 2005 by

**Sage Publications India Pvt Ltd**
B-42, Panchsheel Enclave
New Delhi 110 017
www.indiasage.com

**Sage Publications Inc**
2455 Teller Road
Thousand Oaks, California 91320

**Sage Publications Ltd**
1 Oliver's Yard, 55 City Road
London EC1Y 1SP

Published by Tejeshwar Singh for Sage Publications India Pvt Ltd, typeset by 10/12 Transit521BT in C&M Digital Pvt Ltd, Chennai and printed at Chaman Enterprise, New Delhi.

**Library of Congress Cataloging-in-Publication Data**

vanLoon, Gary W.
  Agricultural sustainability: strategies for assessment / Gary W. vanLoon, S.G. Patil, L.B. Hugar.
    p. cm.
  Includes bibliographical references and index.
  1. Sustainable agriculture. 2. Sustainable development. I. Patil, S.G., 1951-II. Hugar, L.B., 1955-III. Title.

S494.5.S86V25         338.1—dc22         2005         2004027370

**ISBN:**   0-7619-3340-9 (Hb)                81-7829-477-X (India-Hb)

**Sage Production Team:** Vidyadhar Gadgil, Ankush Saikia, Mathew P.J. and Santosh Rawat

# Contents

List of Tables, Figures and Boxes  9
Preface  15

1. **Agricultural Sustainability**  19
   1.1 Sustainable Development  19
       Sustainable development  21
       Background and history  21
       The idea of sustainable development  23
       A controversial concept  26
       Development as improvement  27
       The meaning of sustainable/sustainability  28
       The challenges of working toward
           sustainable development  28
       The sustainability tripod  30
   1.2 Agricultural Sustainability  34
       The sustainability tripod as it applies to agriculture  39
   1.3 Levels of Sustainability  41
   1.4 Studying, Measuring and Assessing
       Agricultural Sustainability  44
       Visions of sustainable agriculture  44
       Building on the vision  45
       Agricultural capital  47
   1.5 Ecosystems  49
       Requirements to maintain an ecosystem  49
       Agricultural ecosystems  52
       Humans and ecosystems  54

2. **Sustainability Indicators**  56
   2.1 Purpose and Properties of Indicators  56
       One indicator or many?  59

|   |   |   |
|---|---|---|
|   | Types of indicators | 60 |
|   | Secondary indicators | 62 |
| 2.2 | Steps in Developing Indicators | 63 |
|   | Conceptual overview and category selection | 63 |
| 2.3 | Desirable Properties of Individual Indicators | 65 |
|   | Relevance, quality and reliability of indicator data | 65 |
|   | Availability and ease of collection of indicator data | 66 |
|   | Stakeholder participation in development | 68 |
|   | Convertibility of indicator data into quantitative terms | 72 |
|   | Integrative capability of the indicator | 76 |
|   | Sensitivity of the indicator to changes over time | 78 |
|   | Ability of the indicator to monitor changes across location and situation | 79 |
|   | Predictive ability of the indicator | 80 |
|   | A clear relationship to vision, action and policy | 82 |
|   | Summary of a general assessment strategy | 83 |
| 2.4 | Quantification and Scaling of Indicators | 84 |
|   | Indicators using the original data | 84 |
|   | Elimination of the scale effect by taking ratios | 86 |
|   | Indicators converted to a common scale (normalising data) | 88 |
|   | Assigning a numerical score to qualitative information | 88 |
|   | Converting various data types to a common scale | 89 |
|   | Selecting values for indicator 'goalposts' | 89 |
|   | Taking weighted averages | 91 |
|   | Inverting data | 93 |
|   | Measuring against 'benchmark' values | 93 |
|   | Combining individual indicators to produce a composite index | 95 |
|   | The human development index | 98 |
|   | Other indicators used to measure human development | 102 |
|   | Diagrammatic ways of combining information | 103 |
| **3.** | **Indicators for Assessing Agricultural Sustainability** | **105** |
| 3.1 | Productivity | 106 |
|   | Productivity potential | 108 |
|   | Productivity indicators | 111 |
|   | Crop productivity indicators | 112 |
|   | Multicropping | 115 |

|     |     |     |
| --- | --- | --- |
|     | Cash crops | 118 |
|     | Monetary indicators of productivity | 119 |
| 3.2 | Stability | 124 |
|     | Soil and water degradation | 127 |
|     | Soil quantity | 128 |
|     | Soil quality | 131 |
|     | Water availability and stability | 144 |
|     | Water quantity | 146 |
|     | Water quality | 149 |
| 3.3 | Efficiency | 153 |
|     | Water use efficiency | 158 |
|     | Energy as an integrative measure of efficiency | 159 |
|     | Energy efficiency calculations | 167 |
|     | Nutrient-based efficiency indicators | 167 |
|     | What should be done with secondary biomass? | 171 |
|     | Biomass as animal fodder | 174 |
|     | Biomass as soil amendment | 174 |
|     | Biomass as fuel | 176 |
|     | Sustainable choices for use of secondary material from agriculture | 181 |
|     | Indicators related to the use of secondary material | 183 |
| 3.4 | Durability | 185 |
|     | Durability in the face of water stress | 187 |
|     | Stress related to pests | 189 |
|     | Economic stress | 194 |
| 3.5 | Compatibility | 196 |
|     | Human (as individuals) compatibility | 197 |
|     | Compatibility with cultural practices | 202 |
|     | Compatibility with the biogeophysical surroundings (Biodiversity) | 203 |
|     | Diversity indicators | 205 |
|     | Other diversity indicators | 210 |
|     | Supporting activities that strengthen the relations between agriculture and the surrounding environment | 211 |
| 3.6 | Equity | 213 |
|     | Equity within an agroecosystem | 214 |
|     | Educational opportunities | 217 |
|     | Income and employment | 219 |
|     | Gender fairness issues | 224 |
|     | Support for human health | 226 |

8  Agricultural Sustainability

|  |  | Food availability and nutrition | 229 |
|---|---|---|---|
|  |  | Response indicators related to equity | 232 |
|  | 3.7 | Evaluation of Agricultural Sustainability Assessment Protocols | 234 |
|  |  | Combining indicator values | 234 |

**4. Studies of Sustainable Agriculture** — 239

    4.1 A Study of Sustainability and Peasant Farming Systems in Zimbabwe — 239
    4.2 A Malaysian Study of Farmer Sustainability — 242
    4.3 Sustainable Land Management in Vietnam, Indonesia and Thailand — 244
    4.4 An Integrated Crop Management Approach to Sustainable Agriculture — 249
    4.5 Indicators for Comparing Performance of Irrigated Agricultural Systems — 251
    4.6 Developing Indicators: Lessons Learned from Central America — 255
    4.7 Sustainability in the Tungabhadra Project Area of South India — 259
        Sustainability of agriculture in the Tungabhadra Project area of Karnataka — 259
        Indicator data from the TBP area — 261
        Policy — 265
    4.8 Assessments of Sustainability over Broad Regions — 266
        Productivity — 268
        Stability — 269
        Efficiency — 270
        Durability — 270
        Compatibility — 270
        Equity — 271

Bibliography — 273
Index — 277
About the Authors — 280

# List of Tables, Figures and Boxes

## Tables

1.1 Strategies for building up various forms of capital required for agricultural food production (based on information contained in Pretty 1999) — 48

2.1 Ecological footprint per capita in selected countries. Data from Wackernagel et al. (2002). — 76
2.2 Carbon dioxide release in selected countries (in 1998) as a pressure indicator of possible climate change effects. Data from the Organisation for Economic Cooperation and Development — 86

3.1 Global long-term average net productivity of the six most important crops. *Primary* refers to the grain or tuber that is the human food product, while *secondary* refers to the remaining above ground parts of the plant. Data from Smil 1993 — 109
3.2 Global, Chinese and Indian average yields for a number of important food crops in 2001. Data taken from the Food and Agriculture Organisation (FAO) database, 2003 — 110
3.3 Qualitative indicators for various crop properties — 115
3.4 Example of yields obtained for two crops grown separately or together. In the multicropping system, the total area is 1 hectare and the ratio of land area for *mung* beans to sorghum is 1:4 — 117
3.5 Rice yields (tonnes per hectare) in a single field between 1976 and 1999 (hypothetical data) — 126
3.6 Visual indicators of soil erosion. Adapted from Stocking and Murnaghan (2001) — 130

| | | |
|---|---|---|
| 3.7 | Indicator Ss1s, describing scores associated with different levels of erosivity that can be observed in the field | 132 |
| 3.8 | Soil physical, chemical and biological property quality parameters | 135 |
| 3.9 | Farmer recommended top ten measures of soil health given in rank order. Paraphrased from Romig et al. (1995), based on interviews with Wisconsin (US) farmers | 137 |
| 3.10 | Irrigation water quality criteria established for use in India | 150 |
| 3.11 | Energy equivalents associated with agricultural activities—chemicals, manufactured materials, human and animal labour, and products | 160 |
| 3.12 | Sample calculation for some aspects of energy efficiency | 168 |
| 3.13 | Sustainable methods of pest and disease control (modified from information obtained from Parrott and Marsden 2002) | 193 |
| 3.14 | A selection of safe drinking water parameters as defined by the World Health Organisation (WHO 2003) | 198 |
| 3.15 | Crop distribution in the head and tail ends of the Tungabhadra Project (Karnataka State, India) command area, in the 1998 and 1999 cropping seasons. From Thompson et al. (2001) | 207 |
| 3.16 | Income distribution for farmers within and adjacent to the Tungabhadra Project (Karnataka State, India) project area. Data for 1999 from vanLoon et al. (2001b) | 223 |
| 4.1 | Some effects and responses of farmers during and after drought in small-scale farming systems in Zimbabwe (summarised from Campbell et al. 1997) | 241 |
| 4.2 | Ratings used to assess practices followed to maintain and enhance soil fertility in cabbage production in Malaysia. Score values can range from −5 to +13, giving an overall distribution of 18 points | 243 |
| 4.3 | Indicators developed in the Framework for Establishing Sustainable Land Management (FESLM) Project in South-East Asia (Lefroy et al. 1999) | 246 |
| 4.4 | Criteria for monitoring agroecological sustainability (summarised from Meerman et al. 1996) | 250 |

| | | |
|---|---|---|
| 4.5 | Sustainability indicators related to agriculture in Central America (Summarised from Segnestam 2000) | 256 |
| 4.6 | Sustainability matrix for the four agroecosystems of the Tungabhadra Project area. Summary (average) of indicator values within the six categories. From vanLoon et al. (2001b) | 264 |
| 4.7 | Levels of agricultural sustainability assessment | 269 |

## Figures

| | | |
|---|---|---|
| 1.1 | The sustainability tripod, showing interrelations between the components of Environment, Economy and Society | 33 |
| 1.2 | Components of the sustainability tripod placed in an hierarchical relationship. Environment holds primacy as the ultimate limiting factor and a vibrant economy is reliant on a 'healthy' society and environment | 34 |
| 1.3 | Global wheat and rice production in the years following the introduction of the Green Revolution (Data obtained from FAO database) | 37 |
| 1.4 | Various levels at which agricultural sustainability can be examined. On larger scales, issues can be termed macro-sustainability, while micro-sustainability refers to issues at a more local level | 42 |
| 1.5 | An ecosystem is a defined area on Earth, subject to inputs from external sources, and outputs into other parts of the environment, but also self-contained in that a number of processes cycle energy, water and nutrients within the system. The result is a relatively stable, interconnected set of species and non-biological components | 51 |
| 1.6 | The distinction between natural and artificial ecosystems is that a purely natural one is maintained by locally-available nature-provided inputs and is maintained in a state that is compatible with its surroundings. An artificial ecosystem requires an augmented supply of inputs beyond those from nature, and is maintained in a state that differs substantially from the equilibrium state that is compatible with the local environment | 53 |

## 12  Agricultural Sustainability

2.1 The components of the DPSIR system of classifying indicators. The example used for each component is that of drinking water quality — 63

2.2 A graphical prediction of the amount of agricultural land (in billion hectares) required to satisfy global food requirements at current and accelerated levels of productivity. Redrawn from Club of Rome (1972: 50) — 74

2.3 Gross codfish catch (in thousand tonnes) in the Canadian Atlantic fishery located in Newfoundland. After the 1991 season (dashed line), the evidence of declining stocks was the impetus for increasing government action to impose strict quotas in the following years (Source: The Department of Fisheries and Oceans, Government of Canada) — 81

2.4 The steps in developing and carrying out an indicator project. Note that the project has a starting point and a conclusion, but in the process there are several opportunities for consultation, feedback and reworking ideas — 83

2.5 Changes in average global temperature between 1860 and 2000 (from IPCC 2001: 107) — 85

2.6 Information triangle showing the flow of information from primary data to final index — 96

2.7 'Diamond' used to illustrate four components of environmental sustainability. Shaded area is a diagrammatic representation of the extent of sustainability for a particular example — 103

3.1 Observable features that indicate soil erosion: (a) Cross-section of a rill. Typical dimensions might be 10 cm width and 5 cm depth. (b) Pedestals are present around obstructions in the soil and indicate that soil has been lost over a broad area, leaving a stabilised portion as a remnant. (c) An armour layer is a layer of coarse aggregates overlying the intact soil, where finer material has been removed — 129

3.2 Relation between resource use and productivity. In this case the resource is water. Productivity of cropping usually increases with greater availability of water but, due to problems of waterlogging and salinity, overuse

|      |                                                                                                                                                                                                                                                                                                                                                                                                    |     |
| ---- | -------------------------------------------------------------------------------------------------------------------------------------------------------------------------------------------------------------------------------------------------------------------------------------------------------------------------------------------------------------------------------------------------- | --- |
|      | of the resource has the potential to severely diminish productivity gains                                                                                                                                                                                                                                                                                                                          | 145 |
| 3.3  | The agricultural process involves a variety of physical inputs, all of which can be described in both energy and monetary terms. Likewise, the products, both primary and secondary, have energy and monetary values. In this diagram, sunlight, water and soil nutrients are considered as natural 'givens', although it is recognised that in irrigated systems water will also have quantifiable energy and financial values | 155 |
| 3.4  | A simplified view of the carbon cycle in an agricultural setting                                                                                                                                                                                                                                                                                                                                   | 177 |
| 3.5  | Production of ethanol from various feedstocks                                                                                                                                                                                                                                                                                                                                                      | 179 |
| 3.6  | Sustainable strategies for use of agricultural 'wastes'                                                                                                                                                                                                                                                                                                                                            | 183 |
| 3.7  | A hierarchy of possible uses of agricultural secondary products ranging from most sustainable uses at the top to least sustainable uses at the bottom. The dividing line separates generally acceptable from generally unacceptable applications                                                                                                                                                   | 184 |
| 3.8  | The nature of the response of an agroecosystem to an imposed stress defines the durability of the system                                                                                                                                                                                                                                                                                           | 187 |
| 3.9  | The figure on the left shows areas of remaining natural vegetation established with no connectedness, while the figure on the right shows the same area of natural vegetation where the existing natural remnants are located in a way that establishes ecological corridors. Shaded areas correspond to various types of native uncultivated vegetation                                            | 210 |
| 3.10 | A canal irrigation system. Within the command area, more water is available at the head end near the reservoir, leading to distribution equity issues that require proper management strategies                                                                                                                                                                                                    | 214 |
| 3.11 | The key features of Enhanced Human Development (equity) based on the UNDP Human Development Report (2002)                                                                                                                                                                                                                                                                                          | 215 |

3.12 Combining individual indicators into category indicators and then into a sustainability index. This diagram is a subset of the diagram shown in Figure 2.6 — 236

4.1 A project-based framework for a study of sustainability of water supplies at the regional level in Central America. Collection of data, development of indicators to measure inputs and outputs lead to improvements in accessibility to water of appropriate quality. (Redrawn in modified form from Segnestam 2000.) — 258

4.2 Levels at which agricultural sustainability can be measured — 267

# Boxes

1.1 Terminology used to describe various types of agricultural practices — 38
1.2 Issues of agricultural sustainability — 43

2.1 Some well-known indicators — 57
2.2 Types of indicators — 61
2.3 A guide to planning gender-sensitive indicators — 69

3.1 Strategy for assessing productivity — 122
3.2 Sustainable methods for maintaining and improving soil fertility — 144
3.3 Strategy for assessing stability — 152
3.4 Strategy for assessing efficiency — 185
3.5 Strategy for assessing durability — 195
3.6 Strategy for assessing compatibility — 212
3.7 Strategy for assessing equity — 233

# Preface

This book is based on ideas developed during the authors' participation in a project to study management factors affecting the sustainability of agriculture in and around an irrigation project area in South India. As we began work on the project, we became aware of the need to work out clear methods for assessing the issues that describe sustainability at a local level. But what are these issues? With this question in mind, we began a search for a comprehensive strategy that would provide the answers we needed. The search involved discussions with farmers in the area where we were working and, at the same time, consultations with experts (largely through published reports, papers and books) who had considered similar questions in other settings and at different times.

The product of these investigations is the subject of this book. We describe an overall approach to assessing agricultural sustainability within a small area, a micro-region. The approach buils on a clear definition of what is meant by sustainability and then sets out six categories that, taken together, provide a comprehensive framework within which individual performance indicators can be constructed. After looking at a methodology for creating useful indicators, we suggest sets of detailed indicators for each category. It is clear that every specific situation will require its own particular method of assessment; as a consequence we emphasise the need to make a selection (from those provided here, and from others that are recommended in a variety of sources) of the most appropriate indicators, and also to be creative in developing new ones. It is our hope that the overall strategies and the detailed examples that we provide will be of use to persons studying, managing and practising agriculture in a variety of circumstances.

The book is particularly directed toward individual farmers, farm managers, project coordinators, planners and agricultural researchers, especially those working in areas where agricultural development and food security issues are subjects of highest priority. Background information is provided, but, most importantly, this is a 'how-to' book, guiding the reader through details of methodology needed to carry out a comprehensive assessment.

The book is divided into four chapters. We begin with a brief discussion of the meanings of sustainability and sustainable development. This leads to a review of the background to this discussion, and of issues related directly to agricultural sustainability. Chapter 2 examines general indicator theory and practice in considerable detail, with emphasis on the desirable properties of indicators and methods used for quantification and aggregation. The need to begin with a well-defined conceptual overview within which categories of assessment are created is also emphasised. The categories that we put forth as necessary for a holistic evaluation of agricultural sustainability are productivity, stability, efficiency, durability, compatibility and equity. Chapter 3 adds detail to the framework by setting out the types of indicators required within each of the six categories. Specific examples are described in some detail, with sample calculations provided in a number of cases. Chapter 4 reviews a variety of situations where others have used different strategies for assessing agricultural sustainability in various settings around the world. While each of the settings is unique, and the assessment strategies are also unique, there are features common to all the assessments. This points to a growing consensus regarding the most appropriate approaches to be used in assessing sustainability of agricultural practices.

It is our hope that this book will contribute to further strengthening this consensus and perhaps to consolidating specific ideas that relate to measuring sustainability.

The book reflects our own opinions that emerge from the extensive fieldwork we have done in the course of this project and over the years in other situations. But, of course, the ideas that we have developed were born and grew within a community of ideas that have evolved through the work of farmers and researchers over many years and in many places. We wish to acknowledge the contributions of all these persons. In particular, we recognise the major contributions to this book of many farmers in the Tungabhadra Project area of Northern Karnataka. They willingly shared information about farming practices and even their individual circumstances—information that enabled us to develop ideas about the essential features of sustainability in communities such as theirs. A number of individuals took part in various aspects of the study. These include N. and M. Pickerl, D. Kulat, G. Mukundarao, J. Yerriswamy, M.S. Veerapur, T. Cross, B. Thompson, K. Sader and J. Calof.

Each of us would like to thank our families for their support during the time we have been involved in the project. In particular, G.W. vanLoon extends his thanks to his wife, Asha vanLoon, who supported the project throughout its duration and also made major contributions in developing methodology, gathering information and assisting with assessment, especially in important areas related to domestic sustainability.

Finally, we thank the Shastri Indo-Canadian Institute, which provided funding and other support to enable the work of this project.

Gary W. vanLoon<br>
S.G. Patil<br>
L.B. Hugar

# 1 Agricultural Sustainability

## 1.1 Sustainable Development

Since the early 1960s food production has increased in many parts of the world. The doubling and tripling of yields of wheat, rice and other crops has been attributed to the 'Green Revolution'—a package of inputs and practices that arose out of research in Mexico, and was first introduced in the late 1950s. New high-yielding varieties and hybrids, use of chemical fertilisers and pest control methods, along with a whole range of carefully-designed management practices, form the core features of Green Revolution technology. With present global annual food production being roughly sufficient to supply an adequate diet to every person in the world, some optimism (and even complacency) has developed concerning our ability to satisfy the basic nutritional needs of the entire human population. But as we go from continent to continent and country to country, distribution of food is far from even. In some parts of the world, diets provide a surplus of high quality (and often high-input) foods. There, farmers are sometimes encouraged not to plant crops because the supply is greater than required. At the same time, in other parts of the world, availability of an adequate, nutritious and attractive diet is far from the norm for a large section of the population. Periodically, droughts and other natural or human activities result in famine and starvation affecting hundreds of thousands, even millions, of people. In some of these situations, for political and logistical reasons we seem unable to solve the distribution problem, and surplus food from other parts of the world never reaches the people in need.

At least as great a problem as the need for equitable distribution is a corresponding need to ensure an abundant food supply not only for the present but also far into the future. This need raises the issue of sustainable agriculture. Investigation into what makes agriculture sustainable is not a new subject. In 1911, Franklin King published a book, *Farmers of Forty Centuries: Permanent Agriculture in China, Korea and Japan* (King 1927), in which he described a labour-intensive system of agriculture that used local resources and relied on recycled village wastes to maintain high yields. He observed that adequate production had continued over four millennia and there was every evidence that productivity could be maintained over further centuries. In other parts of the world as well, farmers have developed systems that will ensure a supply of food both for themselves and for generations to come. At the same time, there is clear evidence of emerging problems associated with many agricultural systems, particularly those that rely on high levels of non-renewable inputs imported into the system.

Problems centre on the uncertainty as to whether productivity can be maintained, but there are social and economic issues as well. Who benefits—the farmer or the consumer? Can small farms be profitable, or is it inevitable that individual farms will grow in size, displacing much of the rural population? And so, in recent years it has become clear that we must focus on questions of food security. Can we maintain and even increase our current ability to produce food indefinitely into the future? Are the practices we have developed within the carrying capacity of the planet, and able to support a good life for a growing population of producers and consumers? To answer these questions, we need to develop methods for evaluating and assessing the sustainable development of agricultural systems, or to put it simply (and for our purposes synonymously), agricultural sustainability. That is the subject of this book.

Because our experience has been largely focused on agriculture in India, our interest is directed towards agriculture in a low-income,[1] tropical situation, where the life of a large section of the population centers on agriculture. The emphasis will be on assessing agriculture on a local scale, within areas sharing common ecological patterns and agricultural practices; nevertheless, the issues concerning sustainable development that we discuss here can also apply, or at least be modified to apply, to other agricultural situations around the world.

## Sustainable development

The two words 'sustainable' and 'development' are loaded with actual meaning as well as many alternate and potential meanings; both words are used and misused in many contexts. Nevertheless, we do not share the popular cynicism that, because they can be so loosely and diversely defined, they have lost any significance. It is our view that there are important, universal and fundamental concepts behind them. The fundamentals form the framework for any definition, yet each situation has its own unique features, and so it is essential to apply the universal concepts to the specific case and to define what is meant by sustainable development. That is what we will do here.

## Background and history

The dual concepts of 'sustainability' and 'development' have been central features defining the course of human history in every part of the world. In the very earliest times, gathering fruits and other plant products along with hunting provided a subsistence diet for primitive societies around the world. Ponting (1991: 21) describes the life of bushmen in South-West Africa as typical of hunting and gathering groups in many places:

> The mainstay of their diet is the highly nutritious mongomongo nut obtained from a drought-resistant tree. It is a very reliable source which keeps for over a year. ... Compared with modern recommended levels of nutrition the diet of the bushmen is more than adequate. ... the amount of effort required to obtain this food is not very great—on average two-and-a-half days a week. The work involved is steady throughout the year (unlike agriculture) and apart from at the height of the dry season the search for food rarely involves travelling more than six miles in a day. Women and men devote about the same amount of time overall to obtaining food but the women, who are responsible for gathering, bring in about twice as much food as the men are able to hunt.

The picture is that here, and in different ways on all the continents, highly stable ways of life based on hunting and gathering were able to persist for millennia. To a large degree, these life-patterns were sustainable because the small populations intuitively recognised that the resources on which human life depends could not be exploited beyond the limits of nature.

This way of life did not however continue indefinitely, in large part due to population pressures. As human populations expanded, more space was required to support the physical needs of each group. As groups began to encroach on each other's land, hunting and gathering as the sole means of obtaining food became an unsustainable practice. During the same period of history, however, there had been 'experiments' in agriculture such as clearing forests to provide space for food species to grow, planting seeds of wild varieties of grain, and selecting those with the most favourable properties. Over time then, the constraints of space and the ability to cultivate crops led to establishment of settled agriculture, which, beginning in various places after 7,000 BCE, slowly emerged as a way of life. This transition could be considered a major development initiated by the human race, and allowed for new ways of life—crops of various types were domesticated, surplus agricultural products could be produced and stored, and structured urban societies were established and grew.

Settled agriculture, by providing surplus food, made it possible for more and more of the population to devote itself to other pursuits. Craftspersons produced pottery, religious artefacts and tools, and classes of people emerged who assigned other activities for themselves. In this way, settled agriculture provided the foundation for many of the subsequent developments in human history in the areas of administration, culture, religion, industry and the military.

The industrial revolution was a second major development affecting human life on the planet. Slowly and throughout the world, innovations in agriculture and industry had been occurring during the era of settled communities. Early in the eighteenth century, however, 'an instance of great change or alteration in affairs' (Landes 1999), which has been termed the industrial revolution, began and expanded in a concentrated manner. This revolution continues into the present and has changed societies in ways that are as fundamental and radical as those that were brought about by the shift from hunting and gathering to settled agriculture. Particularly visible is a move away from rural life to the cities—a process that is nearly complete in many wealthy societies, but is only partial in low-income countries. Superimposed on this massive change has been an exponential growth in population, with resulting issues of food security, use of resources, environmental pollution and the creation and distribution of wealth. These great changes have forced humans to look at questions as to whether life as now lived is truly sustainable.

Coexisting with and informing the physical aspects of life, different religions, cultures and individual human minds in various societies have thought about the place of humans on the Earth and even in the Universe. Some of this thinking has been focussed on issues that describe the relationship between sustainability and development. In a world that can now be described as a global village, we are confronted by glaring evidence of human inequity—extravagant wealth held by one quarter of the global population, with excruciating poverty the lot of another quarter. The need for development leading to peaceful societies and a good standard of living everywhere is obvious to all humane and moral people. At the same time, we are becoming increasingly aware that development in the manner of the wealthy world strains, exploits and contaminates the very resources on which life depends. Is this way of life sustainable? And, fundamentally, can the goals of development and sustainability be reconciled?

## The idea of sustainable development

Thinking about this issue can be traced far back to diverse philosophical and religious ideas emanating from within almost every society. But more recently, particular people and events have brought the subject to the fore. As a single example, in the nineteenth century in California, the building of the Hetch Hetchy dam raised an unprecedented public controversy about the relationship between humans and nature (Jones 1965). Preservationists saw the wilderness as pure and pristine surroundings that should not be spoiled, and within which humans ought to find their place, bearing in mind a central concern to preserve nature's integrity to the fullest extent. This view had roots in European romanticism, expressed in a simple way in books such as *Robinson Crusoe*, but it was also consistent with much of Eastern philosophy, which considered all life as an integrated whole, with humans being but one part of a complex web.

Conservationists, on the other hand, also recognised the importance of values of nature but considered these values in the context of human needs. Nature should be tended and conserved because it served the utilitarian purpose of providing resources for agriculture and industry, and was therefore the basis of human development. Social philosophers who consider human well-being to be the most fundamental of goals tend to hold a conservationist point of view.

What Hetch Hetchy and other events have made explicit is the debate about relations between nature—the setting in which all life struggles and flourishes—and human existence—that special form of life which has built societies whose activities have transformed the Earth. The term 'sustainable development' encompasses the issues that emerge from the tensions in this relationship.

Details of recent attempts to implement the idea of sustainable development and to resolve the conflicts embodied in the term have been provided in several recent books (Chesworth et al. 2002; Damodaran 2001; Dresner 2002; Sachs et al. 1998). A very brief summary is presented here.

The United Nations Conference on the Human Environment held in Stockholm in 1972 was one of the first global meetings that addressed this issue. An important goal of the conference was to examine the ways in which human activities can have an adverse effect on the natural environment. Low-income countries, however, argued that environmental concerns are a luxury of the wealthy and that poverty is the real issue requiring urgent attention. One outcome of the conference was the establishment of the United Nations Environment Programme (UNEP), but the tension between poverty and environment was only made explicit, with no progress toward its resolution. To enable further discussion, in 1983 the United Nations established the World Commission on the Environment and Development, under the leadership of Gro Harlem Brundtland. After many meetings, in 1987 the commission produced a now famous document, *Our Common Future* (WCED 1987), in which a serious attempt was made to examine sustainability and development in a combined way. There was a clear recognition of the scourge of poverty that affects so many in the world and of an urgent need to improve the living standards of the poor. But it was also recognised that poverty alleviation must be achieved in a way that the environment not be degraded, for the sake of future as well as present generations. Reduced consumption by the wealthy and new technology were considered to be two ways in which resources could be conserved and protected, but growth in the global economy was still seen to be inevitable. One of the most often quoted definitions of sustainable development appeared in the Commission's report: '... development which meets the needs of the present without compromising the ability of future generations to meet their own needs.'

But how could this type of development be put into action? The United Nations Conference on Environment and Development held in Rio de Janeiro in 1992 was to be a launching pad for establishing policies and activities directed toward sustainable development, activities in which nations around the world would participate. The environment was at the centre of the conference, and once again impressive documents and directives were put out, most notably a complex plan of action called Agenda 21. After the conference, subsequent meetings and negotiations led to other statements and conventions, including the Convention on Biodiversity and the Kyoto Protocol.

Yet, there has once again been only partial implementation of these ideas by a small number of countries. The tension between the urgent need for development and the difficulties of making development work in a sustainable manner remains as a major impediment to progress in either area. Meanwhile much of the world carries on with its business, with little attention to either issue. Ten years after the Rio conference a follow-up World Summit on Sustainable Development was held in Johannesburg in 2002. While earlier global summits had considered how governments and nations could work toward sustainable development, the emphasis in this conference was on private enterprise as the main engine of progress toward these goals.

Having pointed out the limited concrete results of the global conferences, it remains to be said that there has been at least limited progress toward implementing actions that are consistent with sustainable development. If nothing else, the subject is on the agenda of many countries and some, particularly in Europe, have initiated action in the industrial and agricultural sectors. Furthermore, non-governmental organisations around the world are promoting various causes that respond to the call for improving societies in a sustainable manner. Even many business concerns have seen that it is in both their own as well as the global community's interest that they function in a manner that will ensure their long-term profitability and well-being.

## A controversial concept

Looking at the limited progress in achieving sustainable development, some have argued for abandoning the term. It is pointed out that sustainable development is too vague and can be interpreted in different ways, depending on one's perspective and goals. What does sustainable mean? What is to be sustained and for how long? What is development? For whom? Does it always imply economic growth?

Others point out that the term has been co-opted and used to describe activities that contradict any reasonable meaning of the term. Products have been manufactured and given green-sounding names to indicate a measure of environmental integrity that may or may not have been taken into account in their manufacture. Such practices have been called 'cosmetic environmentalism' or 'greenwashing'. In other cases, governments have also developed programs that carelessly speak of sustainable development, but often incorporate only limited attention to either sustainability or development. This kind of hypocrisy has devalued the term and leads to public cynicism about its validity, even when it is used in a thoughtful way.

Still others note the 'obvious contradiction' in the term sustainable development. Using present statistics for consumption patterns in the high-income world, they point out (correctly) that development up to that level in other countries would far exceed Earth's carrying capacity in terms of resource availability and ability to assimilate wastes.

Each of these arguments tempts us to look for other ways of expressing the complex issues involved. Yet it is our belief that sustainable development is an ideal, a goal that encompasses the best in human thinking about the future of humans on Earth that we all share. It is, in fact, the inherent vagueness that allows parties with different points of view to engage in dialogue. If a rigid definition were imposed, there would be no room for development of common understanding and agreements. Importantly, each specific situation requires examination, and it is not always clear what plan of action fulfils the goals of sustainable development. This will be particularly true as we examine issues of agricultural sustainability. Allowing for different points of view to be expressed in individual cases is essential in the process of expanding understanding and developing best practices.

The practice of greenwashing is lamentable, but it is also a sign of the proponent's need to establish a reputation for sustainability. As the public becomes more sophisticated, attempts to deceive will be challenged, and companies and individuals will be forced to work toward true sustainability.

The apparent contradictions require further discussion and so we will examine both words: sustainable and development.

## Development as improvement

Let us begin with the concept of 'development'. Synonyms for development include a variety of words such as expansion, growth, gain, progress, advancement, improvement, and evolution, and opposites would therefore include contraction, loss, regress, and decline. Many of the synonyms have been a source of problems because of their implied need to call for the use of more space and even greater quantities of the world's limited resources. In the present context, however, we think of development as something closer to evolutionary improvement. In many cases, this can include an element of growth—for agriculture this is often in the form of increased productivity. But growth need not always imply a need for more space or increased use of non-renewable resources. Rather, improvement can also be built on greater efficiency and the closing of natural cycles that have suffered from tangential losses.

Improvement in a social sense is a value-laden concept, and some of our own social value biases will become clear later. Nevertheless there are a number of universal or at least near-universal values—provision of a nutritious diet for all, adequate health care facilities, equitable opportunities for education, etc. It is clear that every society has room for social improvement and so development of the social system is a shared global necessity.

Let us then not think of development as a requirement for some parts of the world and not others. Improvements can be made everywhere, and indeed if we think of the world as a global community, the improvements must have an element of sharing associated with them—so that we can envisage growth in some areas and retrenchment in others, with transfers of physical and social resources back and forth between regions and nations.

## The meaning of sustainable/sustainability

This brings us to the words 'sustainable' and 'sustainability'. We all have a general sense of what is implied by these two words. A sustainable practice is one that can continue indefinitely. In an agricultural system, sustainability essentially means that crop production and economic gain will flourish over a very long, essentially infinite period of time. But what is required to allow productivity to continue? Does maintenance of the productive practices require the importation of large quantities of resources from elsewhere? If so, it may be sustainable (at least in the short term) on a local scale, but problems arise when one broadens the question to include the global picture. Indeed, sustainability is ultimately a concept that can only be considered in the global context. Having said this, we must emphasise again that this book will be concerned with a much more local scale, but also keeping the broader issues in view.

On another front, does maintaining the physical success of an enterprise over the long term require repressive social practices? It would require a very limited view of sustainability to consider such activities truly sustainable.

Science and technology clearly relate to many aspects of sustainability. Issues of sustainability are however not merely technical problems; rather, thinking about them must be constructed on a broad-based worldview that cuts across individual disciplines. Specific scientific information and knowledge is essential to inform many aspects of the subject, but our understanding of science is itself evolving and issues are rarely clear-cut. And importantly, the issues go far beyond science in the commonly accepted sense of something that can ultimately be 'known', after which 'correct action' can be taken. Sustainability is an all-encompassing vision of what life ought to be, where all the physical and social sciences, religion, philosophy and ethics come together (for a comprehensive view of the various issues related to sustainable development and its assessment in a tropical country, see Hall 2000).

## The challenges of working toward sustainable development

As has frequently been noted, there is an inherent and inevitable tension in linking the two components of our theme—*sustainability* and *development*. We cannot deny that there may be difficulties, even contradictions, encountered when one attempts to reconcile these two issues. Take the case of forestry. Clear-cutting a well-established mixed forest may sometimes be justified because, through the efficient methods that this enables, it can be highly productive and bring economic and even social benefits to a community. But a forest management strategy based on clear-cutting can, at the same time, have serious adverse environmental consequences. Similar is the case of fisheries. Drag-net fishing by deep-sea trawlers could have enormous immediate and short-term economic benefits for a coastal community, but it could be (and has in some instances been proven to be) completely unsustainable in terms of preserving the resource.

Part of the solution to these conundrums is to consider every issue from a long-term perspective. Many apparent economic or social benefits of development are short-term and clearly cannot pass a test of sustainability, as is the case for the two examples given here. Further, one of the key features of sustainable development is the future perspective, with a goal of leaving a world whose physical, biological and social resources can satisfy the needs of future generations.

Taking the long-term view is not sufficient, however, as a solution to all problems where the diverse needs of development come into conflict. The long-term view should not become a means by which we put aside concerns about present needs. Severing the inhabitants of forest or fishing communities from their means of livelihood without provision for acceptable alternative working and living conditions is equally an unsustainable solution.

Where these difficult situations exist, implementing sustainable development practices is an intrinsically challenging activity and one that is not without costs. True sustainability is an ideal that requires creative thinking by all the stakeholders in every situation. Implementation of any course of action must involve participation of the global community, not just those who may be immediately benefited or disadvantaged by a particular decision. In the forestry example, resolution of conflicting priorities might involve selective tree cutting, while ensuring that new growth takes place where trees have been removed. This is, of course, a more labour-intensive operation, a fact that would be reflected in the need for a greater number of well-trained forestry workers with the consequent higher production costs. It is in the resultant higher prices that the broader community, the consumers of forest products, would be called upon to contribute to what are clearly more sustainable practices.

In a similar manner, as we examine issues of agricultural sustainability at the farm and community level, we must be aware that actions taken on the ground need to be complemented by supportive actions at higher levels. These actions may include a whole range of policies concerning regulation and pricing of agricultural inputs, provision of irrigation water, land tenure, social services, control of commodity prices and global trade issues.

The apparent contradictions between *developing* and *being sustainable* at one and the same time also come into focus if development carries with it a sense of growth or physical improvement. Who can deny that a person living on a meagre diet that provides only 1,600 calories a day should not aspire to and deserve the required complement of around 2,700 calories? This means around 140 kg additional grain every year for that single person. Or who can say that 15 GJ (gigajoules or billion joules)[2] of energy per person to satisfy all needs—cooking, lighting, heating, transport, agricultural work—is sufficient for comfortable living? This is the present annual per capita amount consumption of the 4.2 million citizens of Sierra Leone. It has been suggested that a modest requirement would be closer to 50 GJ per capita per year. Development in these situations must surely mean growth in the energy sector. And is it not possible that development must also mean contraction of resource use—for example, for citizens of North America who consume about 300 GJ per capita annually, and many of whom live on extravagant diets with excessive protein?

The human challenge of achieving the goal of sustainable development is the need to reconcile that which is sometimes irreconcilable.

## The sustainability tripod

A comprehensive view of what is sustainable must therefore involve many factors, and there is broad consensus that these factors should incorporate three elements: environmental, economic, and social.

### Environment

It was issues surrounding the environment and ecology that brought modern concepts of sustainability to the fore. One of these ideas is the recognition of the connectedness of all of nature. In many cultures, this concept has been the basis of life experience and philosophy since time immemorial. In others, there has been a tendency to set human experience apart from its natural surroundings, or at least to see humans as having the ability and obligation to control nature for their own benefit (see, for example, White 1967).

In recent years, these disparate views have come closer together. Beginning in the nineteenth century, but at an accelerating pace in the last half of the twentieth century, biologists began examining and expressing the relations between humans and their environment in more scientific terms. Through early concepts of ecology came ideas and specific knowledge that demonstrated how all forms of life are intricately interwoven with each other as well as with the inanimate world, and how manipulations in one area can have a cascading effect on processes and populations seemingly far removed from where the manipulation occurred. Studies on pollution have illustrated the fragility of the environment. Cases have been described where human activities have caused irreparable damage to land or water bodies. Studies of resource availability teach us that the supplies of minerals and fuels that we extract from the Earth are not inexhaustible. All these issues have come together, and the world community is now increasingly aware of the uncertainty regarding whether human activity can continue to operate within a pattern that exploits rather than cooperates with nature. For this reason, there is growing awareness of the centrality of environment to issues of sustainability.

## Economy

Economic issues have also been prominent in developing modern ideas of sustainability. Notwithstanding the views of some radical idealists, economics governs many aspects of the day-to-day life of nations and of most individuals who make up the world's population, and the core features of every economic activity come down to costs and benefits. This is not to say that assessment of either costs or benefits is a simple matter. The well-known measure of prosperity, the Gross National Product (GNP), calculated on a per capita basis, is a case in point. In a broad way it measures the economic value (benefit) of all goods and services produced within a nation. But it measures economic activity irrespective of its value to society, and does not take into account side issues (externalities) that, while critical, may not easily be assigned a monetary value. We will say more about these kinds of situations later.

In spite of difficulties in assigning costs and benefits, some attempts to do this have been made and are essential.[3] After all, the industrial, agricultural, cultural and service enterprises in which humans are involved include costs like time, use of space and resources, and money, as well as benefits in terms of creating employment and the provision of services, goods and profit. It follows then that a sustainable practice must be one that can be justified in broad economic terms, but these economic terms must be evaluated within the context of environmental constraints and with the needs of society always at the forefront.

## Society

We have stated that there is a broad consensus regarding a tripod of features related to sustainability. This may be so, but it is not without controversy that social aspects have been included as one of the legs of the sustainability tripod. Some people share the concept that social issues, important though they may be, should be treated as a separate matter. After all, social values change and so a decision to choose a particular value and demand that sustainability should maintain or enhance that value may be questioned and require modification at a later time. Indeed, many decisions regarding the nature of social goods are based on culture and religion, and human history shows us that these vary over time as well as from place to place around the globe. So, it is argued, we can never come to a common agreement about specific social values that we all share and that should be maintained.

Nevertheless, as we have stated earlier, we believe that at least some social values have gained universal or near-universal acceptance. These are the broad principles of equitable access to resources as support for a decent standard of living, opportunities to participate in decision-making, access to education that provides opportunities for cultural and intellectual growth, and provision for adequate healthcare. The social leg of the sustainability tripod is constructed from these ideas because no human activity can continue and flourish unless it fully incorporates individual and social needs.

Figure 1.1 illustrates the three components of the sustainability tripod and shows that each component is interconnected with the others. It is only when we consider all three components together as a group, that we have assessed sustainability in comprehensive terms.

**Figure 1.1**
The sustainability tripod, showing interrelations between the components of Environment, Economy and Society

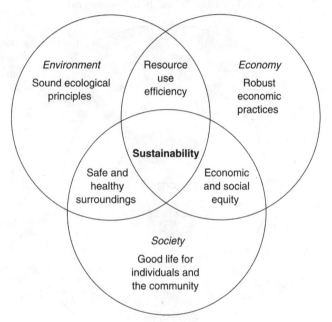

There are alternative diagrammatic representations that attempt to illustrate other aspects of the relation between the three essential components of sustainability. Figure 1.2 describes sustainability in terms of concentric circles, where the environment is the ultimate setting within which societal structures are built, and society itself is more fundamental than the economic constructions that humans design and implement. Yet, in this representation too there are interconnections, as shown by the double-headed arrows between each of the three components of sustainability. Most important here is the emphasis on the primacy of the environment. While economic and social systems evolve and change—sometimes with the development of organisational structures that fail and are replaced—every society and system is built and lives within the diverse surroundings of the Earth. Every human-created form of sustainable society requires that

**Figure 1.2**
Components of the sustainability tripod placed in an hierarchical relationship. Environment holds primacy as the ultimate limiting factor and a vibrant economy is reliant on a 'healthy' society and environment

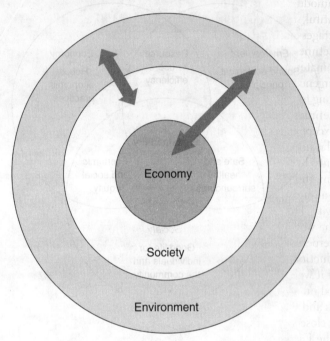

the integrity of the environment be upheld. Should our surroundings be altered in a way that life itself is endangered, no social or economic manipulations would suffice to revive sustainable life. Perhaps this self-evident but often ignored principle is most clearly exemplified in the agricultural enterprise!

## 1.2 Agricultural Sustainability

In terms of the development of nations and their prosperity, public attention and the focus of the media are often directed toward urban issues: industry, commerce, communications, culture and recreation.

Discussions regarding the subject of trade frequently centre on major commodities such as petroleum and other natural resources, as well as manufactured goods. Within this context, food is treated as just another commodity and it is sometimes assumed that there will always be a plentiful supply. Agriculture becomes a topic of discussion only when shortages occur. Only when failing rains lead to food shortages and sometimes famine and displacement does food become an item that dominates world news, at least until the problem is 'solved', often by emergency shipments of food from distant sources. Cyclical factors leading to high prices for luxury items like coffee is another subject that sometimes stirs general interest in agriculture in the wealthy nations.

Except in these situations, well-to-do people frequently assume that food is not really a problem, and that shortages can be overcome in perpetuity by new scientific advances. In other words, without thinking the matter through, we have decided that agriculture as now practiced will always be able to feed the world, and is therefore a sustainable activity. This complacent attitude is not, however, shared by everyone in the world, especially not by the millions who live under conditions of perpetual food deficit. It may be true that the present global food production rate is approximately sufficient to satisfy the basic needs of all if it were distributed in an equitable manner. But this conclusion is based on an average value that hides major unevenness in yearly harvests and distribution around the Earth. The actual situation deserves very close scrutiny, and we will summarise some of the key issues here.

The Earth's total population in 2004 was approximately 6.4 billion persons; every four days about one million persons are added to this total, for an annual increase of 90 million persons, or one billion persons every 11 years. This number is not evenly distributed across the countries of the world. The population density ranges from less than 10 persons per square kilometre in some countries that are located mostly in temperate regions (countries like Canada and Australia) to more than 200 persons per square kilometre in many tropical countries. The finite global environment must support this large, unevenly distributed number of people. Human life has many essential requirements, and perhaps the most fundamental physical needs are air, water and food. It is the last of these items that is our concern as we consider the subject of agriculture. But we shall see also that there is a relation between agriculture and the availability and quality of the other two fundamental elements, air and water. In a very real sense, none of these issues can be examined in isolation.

While a proper human diet requires that there be a balanced complement of nutrients—proteins, minerals, vitamins, etc.—the most basic need is for an adequate supply of energy, usually measured in terms of calories. Energy is provided by consumption of carbohydrates, proteins and fats in foods, and for many people the basic energy source is some sort of cereal (wheat, rice, sorghum, maize, etc.) or carbohydrate-rich roots like cassava or yam. The average human being requires a daily intake of food with an energy content of approximately 2,700 food calories[4] (equivalent to 11 million joules). Most cereals have an energy content of about 14 million joules per kilogram of grain. Therefore, we can estimate that the basic dietary need is for food providing energy equivalent to 0.8 kilograms of grain a day. This corresponds to 290 kg a year per person, or 1,800 million tonnes (grain equivalent) for the total world population of 6.4 billion.

Of course, 'man does not live on bread alone'; especially in high-income countries the diet is highly varied and there are many other sources of the needed food energy. However the animal sources, which contribute in such a major way to much of the more wealthy world's diet, are themselves also based on grain. Depending on how the animals are raised, it may require between 3 and 13 kg of grain to produce 1 kg of meat.

Making these various assumptions, the world's cereal harvest can be compared with the global need for food energy. In the year 2001, the global production of cereals was 2,086 million tonnes ($2.9 \times 10^{19}$ J), just about enough to supply an adequate diet (3,000 cal per day) for every person. Distribution of this grain supply is, however, far from equitable.

In the wealthy nations, many persons (some estimates put the number at about 15 per cent of the world's population), including a good number whose sedentary occupations and lifestyles involve only light physical activity, have a daily intake of 3,500 cal or more and obesity is a growing problem. Furthermore, much of the food is in the form of animal products, meaning that the actual energy content is even greater. At the same time, another 15 per cent of the world's population is undernourished, with many people subsisting on a diet supplying less than 2,000 calories, and subject to other serious nutritional deficiencies as well. While the global average may suggest an adequate food supply, the average does not inform us about inequities arising from unequal distribution of food.

Can the world be fed in the future, when we consider the growing population? The population has been increasing at a rapid rate over

**Figure 1.3**
Global wheat and rice production in the years following the introduction of the Green Resolution

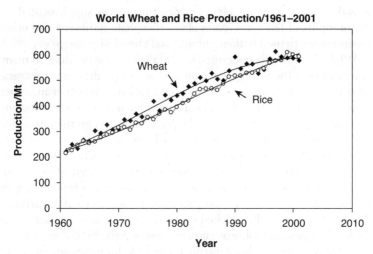

Source: FAOSTAT data (2004), at http://faostat.fao.org.

the years. But fortunately, the rate of food production has also gone up (see Figure 1.3). Just to maintain the steady state as we move into the future, each year the 90 million additional people will require 26 million tonnes (as grain equivalents) of food.

Over the past 40 years, global annual production of both wheat and rice has increased by a yearly average of about 9 million tonnes. Production of other staple foods has also grown in a similar fashion, giving a combined increase of 29 million tonnes of total cereals per year. The rate of increasing production and growing need have therefore been in balance over the past four decades.

The impressive gains in food production are a consequence of several factors. For one, some marginal land, formerly unused or underused for agriculture has been taken up for cultivation. There is very little of this resource available for future expansion, and converting land such as that which supports tropical rain forests into farmland can have serious environmental consequences for the Earth as a whole. Forests including those in the Amazon valley, central Africa and parts of Indonesia have been described as 'the lungs of the Earth' and they are home to a diverse array of plant, animal and microorganism species found nowhere else on the planet. Furthermore, the soils that support this lush and varied growth are also not suitable for agriculture unless supplemented by major inputs of fertilisers (particularly phosphate fertilisers) and other amendments to modify their physical and chemical properties.

While there is little underused land available for agriculture, in many places all over the world significant portions of productive agricultural land are being lost due to erosion, desertification, salinisation and other causes. In spite of this, the global foodgrain production has steadily increased over the past 40 years. The principal reason for the impressive gains in food production, however, has been the introduction of new varieties of grains and the associated management practices that accompany the new strains. A major component of this so-called 'green revolution' has been the requirement for a package of inputs beginning with the new seeds but also including an assured supply of water, adequate plant nutrition (usually supplied via inorganic chemical fertilisers), and methods of pest control, again usually involving synthetic chemicals.

These practices are substantially responsible for increasing yield and allowing food production to keep pace with population growth up to the present. While global food production appears adequate at present, there are many causes for concern as we look to the future. Degradation of good agricultural land, conversion of forest and other unique habitats to agriculture, and reliance on chemical and other non-renewable inputs in the long term could have a negative impact on food security and on rural life in general. Even today, there is growing evidence that current practices are problematic in terms of long-term sustainability.

> **Box 1.1**
> **Terminology used to describe various types of agricultural practices**
>
> - Low-input agriculture—Agricultural practices that attempt to minimise the use of non-renewable inputs, including machine energy
> - Organic agriculture—Agricultural practices that eliminate all chemical inputs, including synthetic fertilisers and pesticides

*(Continued)*

*(Continued)*

- Ecological agriculture—Agricultural practices that emphasise the relationship with the surroundings, and take measures to maintain the integrity of the local environment
- Alternative agriculture—A vaguely defined term that can be used to describe any agricultural practices that incorporate some selection of practices included in the definitions above

All of the above terms focus on the biological and physical aspects of agricultural practices; sustainable agricultural practices must include consideration of social and economic issues as well as those related to the biophysical environment.

## The sustainability tripod as it applies to agriculture

### Agricultural sustainability: Environment

While we often associate environmental problems—pollution, greenhouse gas emissions, loss of habitat—to industrial development, agriculture is in fact itself a major contributor to a whole range of environmental problems. Even in parts of the world where industrialisation is not highly developed, environmental degradation can be very severe, mostly associated with the diverse activities involved in agriculture and rural domestic life. The list of agriculture-based environmental problems is a long one: loss of biodiversity and destruction of natural habitats, over-consumption of surface water and groundwater, contamination of soil and water by organic biocides, nutrient-induced eutrophication of water bodies, microbial and nitrate contamination of drinking water supplies, and release of excessive quantities of the greenhouse gases such as carbon dioxide, methane and nitrous oxide, associated with specific agricultural practices.

On the other hand, it is equally possible that carefully executed agricultural planning and practice can enhance environmental values by stimulating growth of a diverse variety of crops, incorporating animals into the web of activities, using and reusing the products, and ensuring that the surroundings remain in as natural a state as possible, untouched by release of undesirable and toxic chemicals and microorganisms.

### Agricultural sustainability: Economy

The economic element of agriculture relates to individuals as well as to the local community and the broader national society. In being a producer of food, each farmer carries out agriculture, first of all, in order to be able to provide for the physical needs of his/her own family, and paramount among these is the need for food. If the system does not make economic sense, if farming is not profitable, and if basic needs are not provided for, the system is unsustainable. Society at large also requires that sustainable practices have a sound economic basis. Long-term subsidies to large groups of either producers or consumers distort the true picture of costs and can only be maintained by support from the profits of other sectors. Agricultural sustainability then demands that farmers continue to make a good living and that the population as a whole be supplied with an abundance of high quality food at reasonable cost.

### Agricultural sustainability: Society

Whether an agricultural system is sustainable or not goes beyond evaluation of the physical-biological processes and the economics of the situation. More than in most other occupations, agriculture is a way of life. Many people are born into this way of life and/or choose it in preference to what others may describe as the advantages of life in the city. Sustaining rural life must then provide the basic services related to education, health, recreation, etc., that all humans have a right to expect. But it must provide more—especially opportunities to be challenged and to learn and contribute to improvements in the world around them.

The social element of the sustainability tripod derives also from the fact that each one of us is involved directly and indirectly in all aspects of agriculture. Direct involvement ranges from 2–3 per cent of the population in many of the industrialised northern economies to the vast majority of the population (70 per cent or more) in some countries of Africa, Asia and South America. Indirectly, we are all involved as consumers of food and fibre products, although it is an unfortunate fact that many of us live lives detached and distanced from agricultural processes. Agriculture is also a time- and labour-intensive occupation. Where the resource base is deficient, the demands for long hours and heavy physical work can be especially severe. It is therefore an essential requirement for human well-being that requirements for being a successful farmer not be so harsh as to preclude opportunities for education, recreation and relaxation. This applies to all the members of the farm family as well as the rural community as a whole. In some countries, government policy has encouraged productivity at the expense of other factors, including a need to provide facilities and opportunities for all in the rural community. One consequence is consolidation of land and resources in the hands of a small proportion of the total population, leaving many people with marginal holdings or even no land at all. Unless adequate support by way of accessible employment in other activities is available, an unsustainable population of dispossessed persons develops in the rural community.

It is beyond question that agriculture is one of the most fundamental and essential of all human activities and that unless it continues to flourish, the world as we know it cannot exist. As is the case for sustainability in any setting, sustainable agriculture must consider and bring together sound practices in the environmental, economic and social spheres—the three legs of the sustainability tripod.

## 1.3 Levels of Sustainability

Issues of agricultural sustainability are subjects that can be considered across a spectrum of levels—from a global scale to the scale of the farm field, and at every level between (see Figure 1.4).

On a global scale, the focus may be on the world food supply, maintenance of adequate arable land, regulations for trade in agricultural commodities and impacts of agriculture on global climate. Continents and nations share similar concerns. Trade policy related to agricultural commodities is an ongoing subject for discussion, and it directly impacts on national policies regarding agriculture in general. Agricultural issues are frequently studied on the scale of regions, like the great plains of North America or the highlands of East Africa, where common ecological features are shared. Such regions are sometimes restricted within the boundaries of a single nation, but more often straddle boundaries and cover a major portion of a continent. Watersheds, both natural and artificial, form the basis for important aspects of rural planning, and the hydrological cycle within a watershed plays a central role in agriculture. Because activities in the upstream portion of the watershed frequently affect

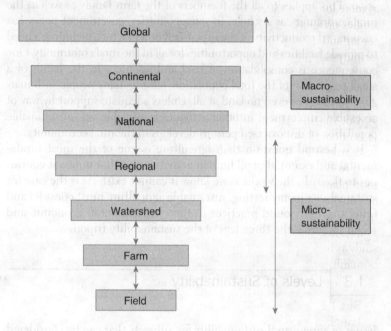

**Figure 1.4**
Various levels at which agricultural sustainability can be examined. On larger scales, issues can be termed macro-sustainability, while micro-sustainability refers to issues at a more local level

environmental processes downstream, it is important to account for this in developing local policies concerned with agricultural development. At the small-scale end of the spectrum of levels of sustainability are the farm and the individual field. Here, farmers' concerns centre on crop productivity, provision of sufficient profit to support individual families, and quality of life for those whose lives depend on that field and ones nearby.

The hierarchy of levels of sustainability is meant to show several things. All of the levels are interconnected, and although we have used the term 'hierarchy', this does not suggest any order of precedence or importance. The double-headed arrows point to the fact that information, policies and actions associated with each stage affect sustainability issues at every other level above and below. Supra-national policies are determined and influenced by those of individual governments and these, in turn, influence decisions at every level below. Likewise, what is achieved on individual fields and farms cascades upwards, affecting the status of sustainability at higher levels.

In terms of sustainability assessment, too, each level is interconnected. Data concerning the world at large, and about nations and vast ecologically connected regions, all provide a background and context for studies at more local levels. And the many pieces of information obtained from studies on small plots of land, on farms and in individual villages, add colour and detail that builds up the larger pictures. Roughly speaking, for purposes of assessing agricultural sustainability in the larger regions, much data have been collected and are available from major governmental and other scientific, economic and social organisations. These data, gathered for a variety of purposes, can be used to follow trends that provide valuable information about sustainability. Using the terminology of economics, we can generally speak of work in this area as a macro-sustainability study. As we approach more local regions, there is usually a need to collect highly specific information, and the individual agronomist, planner or social scientist, along with the farmers themselves, will be personally and directly involved in the data collection process. Even here, however, concepts and information from the more broad-based sources will also be incorporated in the work. We refer to this type of study as being in the field of micro-sustainability. This book is concerned primarily with measurements of sustainability at the micro level.

---

**Box 1.2**
**Issues of agricultural sustainability**

**Macro-sustainability issues related to agriculture**

- National and global resource consumption
- Greenhouse gas production/sequestering
- International trade—monetary and environmental regulations
- Loss of genetic diversity—legislation to regulate
- Equity in food supplies between nations

**Micro-sustainability issues related to agriculture**

- Productivity of individual farmers
- Availability of financial and physical resources
- Financial viability to farmers
- Ability to grow crops in a safe manner
- Equity within the local and national community

## 1.4 Studying, Measuring and Assessing Agricultural Sustainability

### Visions of sustainable agriculture

In any assessment of sustainability, an essential first step is to set out a broad vision of what is meant by sustainability *in that situation*. Only after establishing such a clear overview can a coherent plan be made, as the basis for addressing individual questions.

In many different contexts, a number of broad concepts or vision statements of what is meant by sustainable agriculture have been offered.

As early as the first century BCE, Marcus Terenius Varro, a Roman landowner, defined sustainability by recording in his treatise, *Rerum Rusticarum*, a view that agriculture is 'a science, which teaches us what crops are to be planted in each kind of soil, and what operations are to be carried on, in order that the land may produce the highest yields in perpetuity' (quoted in Conway 1997).

In recent years, a variety of more modern, but in many ways similar, descriptive statements have been proposed as definitions for agricultural sustainability:

> ... the ability of an agroecosystem to maintain production through time, in the face of long-term ecological constraints and socio-economic pressures (Altieri 1987).

> ... the condition of being able to harvest biomass from a system in perpetuity because the ability of the system to renew itself or be renewed is not compromised (Gliessmann 1998).

> Sustainable agriculture involves a system for food and fibre production that can maintain high levels of production with minimal environmental impact and can support viable rural communities (Mellon et al. 1995).

For the purposes of this book, we think of agricultural sustainability as an essential human activity—in fact, as one of the most fundamental of human activities. It can be a necessity for an individual family, in order to supply itself with food as well as income from employment. It is equally a necessity for society as a whole, in order to provide food and other agricultural products for all, including those who are not themselves engaged in agriculture. With this in mind we will build a strategy for assessment of agricultural sustainability around the following vision:

> Sustainable agriculture is the activity of growing food and fibre in a productive and economically efficient manner, using practices that maintain or enhance the quality of the local and surrounding environment—soil, water, air and all living things. It is also sustainable in supporting the health and quality of life of individual farmers, their families and the community as a whole.

## Building on the vision

Starting with this broad definition for sustainability in agriculture and rural life, many follow-up questions can be asked in order to develop a useful strategy for assessment. Are we looking to maximise use of land for agriculture or, alternatively, to return large areas of land to a 'wild state'? Does sustainability require plans to minimise human labour or to provide enhanced opportunities for rural employment? Is land to be owned collectively or individually, and in the latter situation is there a place for non-landowning farm labourers? These and many other questions demand examination and decisions that are explicit and clear.

Sometimes difficult choices are required; there are no absolutes and the rationale supporting the selected option must then be explained. Essentially the choices are at one and the same time individual, social and political. For some issues, there may be broad consensus supporting what is chosen, ensuring general acceptance and limited controversy. With other issues, there is a wide diversity of opinions and this underlines the need to ensure full participation of every group in the population in all phases of development and measurement of sustainability.

We might assume that controversy will occur more frequently when one is dealing with the social components of sustainability. It is true that controversy can occur in this realm. What is equity in terms of present and future generations, in terms of gender, in terms of class? What is required to ensure that individual opinions are heard and respected? How do you provide adequate opportunities for personal fulfilment for all members of a society?

Recent history in China provides an excellent example of changing concepts regarding agriculture and rural life. In its initial stages, the 1948 revolution defined rural society in terms of redistribution of land according to egalitarian principles. The goals were to support villages where individuals would work their own plots, producing grain and other crops and livestock for their personal use, for the village and for the wider society. During the early 1950s and up to 1962, policies changed and there was a push toward collectivisation, wherein most of the productive activities would be carried out within larger units: collectives and then communes that could involve a number of villages and thousands of people. At its height, the collective push led to sharing of most actives of daily living: organised field work, communal meals and childcare. This approach did not, however, remain as the politically favoured means, and in subsequent years agriculture and village life has reverted to a more individual family farm approach. Along with the changing socio-economic patterns of food production, views regarding use of resources have also evolved. At one point, directives from the central government sought maximisation of land for agricultural purposes, so trees and forests were cut down in favour of conversion of land for agronomy and horticulture. Soil erosion, reduced rainfall, encroaching desertification in some parts of the country and loss of other benefits from having forests were among the causes for a reversal of this decision. In the early twenty-first century there is renewed emphasis on the virtues of planting trees and a countrywide reforestation programme has been initiated.

Within these diverse ways of thinking, what would a comprehensive study of sustainability have revealed? Defining the 'terms of reference' is an essential starting point, but a well-defined examination could itself become a critic of the concept. In other words, we cannot assume that every defined worldview is a potentially sustainable worldview.

Policy issues related to economics and society, such as those affecting rural life in China, are well-known subjects of controversy. Yet we should not have the impression that scientific issues are based on 'fact' and are above controversy. Especially in areas of environmental science, including in applications of science to agriculture, there are evolving and controversial views of what is correct or what is best in a particular situation. Changing views about the safety and applicability of pesticides is one example, as are issues surrounding the subject of global climate change. Information about greenhouse gases, some generated from agricultural activities, has become available only in recent years, and the significance of these recent findings is a subject of ongoing discussion. Debates about the role of genetically modified organisms (GMOs) in sustainable agriculture have also begun only recently and will doubtless continue for some years. Strong arguments can be put forth concerning their contribution to environmental improvement in that they can potentially reduce the need for chemicals to control insect and weed infestation. Equally strong arguments point to concerns about crops themselves becoming weeds, and about gene transfers into wild species, creating new species of uncontrollable 'superweeds'. This debate (and there are many other aspects to it) is like sustainability itself. Which side is right can only be determined over the long term, and decisions are required in the present.

The point of this discussion is that each study of sustainability should begin by stating and justifying the axioms or parameters on which the detailed work plan is based. This will form the conceptual overview that informs the questions and evaluations that follow. In other words, the starting point is: 'What do you want to sustain, and what are acceptable parameters for its sustenance?'

Therefore, in a process of measuring sustainability we must prepare ourselves for a dialogue beginning with a definition of ideals, proceeding to an examination of the particular situation with respect to these ideals. The examination will show how close or far the system is to functioning sustainably within the defined context. This critique may then demand a re-examination of the originally defined overview, and so the back-and-forth process could continue. This iteration can itself be a way of approaching an appropriate working definition for sustainability within accepted norms.

## Agricultural capital

There will be different views about the specific components of what is necessary to ensure sustainability within a given situation. In every case, however, agriculture relies on available resources of various kinds. Often referred to as 'capital', these resources include:

**Table 1.1**
Strategies for building up various forms of capital required for agricultural food production

| | |
|---|---|
| *Natural capital* | *Financial capital* |
| • Water harvesting, water management | • Stable markets |
| • Soil conservation | • Subsidiary activities |
| • Biological pest control | • Readily available credit |
| • Composting, manuring | • Post-harvest technological opportunities |
| • Diverse systems—many types | • Value-added activities |
| • Conserving genetic resources | |
| | *Physical capital* |
| *Social capital* | • Improved tools, machinery |
| • Cooperatives | • Precision agriculture methods |
| • Extension workers: Government, NGO, private | • Low dose sprays |
| • Farmers self-help and research activities | • Improved crop varieties |
| • Social values and systems | • Roads |
| | • Processing plant |
| *Human capital* | |
| • Improved nutrition | |
| • Education | |
| • Health | |

Source: Based on information contained in Pretty (1999).

- Natural capital—the soil resource, water from rainfall or other sources, the air, animals used for their labour and as a source of manure, the surrounding natural vegetation
- Human capital—humans who supply labour, not only physical labour but also intellectual input for planning production strategies
- Social capital—systems providing labour and marketing support as well as information related to agriculture and health services
- Financial capital—markets for purchase and sale of goods, a credit system supplying funds to all levels of agricultural workers
- Physical capital—implements needed for agriculture, roads and means of transport, factories for processing of farm produce

Sustainable agricultural practices take into account all these resources. Clearly, the various forms of capital are continuously being used in the processes of food production. At the same time, operations are required that are sensitive to the need to *build up* as well as to *consume* capital of every kind, so that a balance is maintained. One study (Pretty 1999) documents some of the practices that are consistent with the building up of capital reserves of various kinds (see Table 1.1). It is evident that in various ways this set of practices relates to the three sustainability components.

## 1.5 Ecosystems

The general concepts of ecology are essential background for assessing agricultural systems. To start with, we should develop a clear idea about the meaning of the term 'ecosystem'.

An ecosystem is defined as a recognisable, defined area of the Earth, sharing common structural features and being maintained by specific interrelated biotic and abiotic processes. Ecosystems may be large, such as a vast expanse of desert, or very small, like a pond or small forest. Most importantly, within an ecosystem the living and non-living components work together to establish relatively stable and predictable features that make it possible to identify that area.

There are both external and internal aspects to an ecosystem.

### Requirements to maintain an ecosystem

In order to maintain an ecosystem in its relatively stable state, it is necessary to have access to an assured supply of three essential external components, energy, water and nutrients. Energy is a fundamental requirement for all life and ultimately most of the Earth's energy comes from the Sun. Besides providing heat, solar radiant energy is the driving force that enables photosynthesis to take place. Photosynthesis is the process through which plants and algae (in biological terms these are called producers or autotrophs) grow by assimilating carbon dioxide and water to form energy-rich carbon compounds. Carbon dioxide and water are found in the atmosphere in gaseous form as well as in soils or water bodies; depending on environmental conditions the needed carbon dioxide is present in the form of several different carbonate species. A simple depiction of the chemical reaction describing the photosynthetic process in terrestrial plants is given here:

$$\text{carbon dioxide} + \text{water} \rightarrow \text{organic matter}^5 + \text{oxygen}$$

$$CO_2 + H_2O \rightarrow \{CH_2O\} + O_2$$

Of course, plants and algae require more than carbon dioxide and water in order to develop and mature, and the organic compounds formed also contain elements other than carbon, hydrogen and oxygen. Additional nutrients are essential for biomass to grow, with nitrogen, phosphorus and potassium being required in the largest quantities, along with a number of other minor and trace elements. The nutrients allow for the synthesis of complex biological components that serve special structural and metabolic functions. Using solar energy, carbon dioxide, water and nutrients, the many and varied forms of plant life are produced—each with distinctive features, but all sharing common properties that enable us to refer to them in a general way as biomass.

Biomass becomes food for heterotrophic organisms (also called decomposers). Decomposers include microorganisms like fungi that feed on decaying organic matter, in the process incorporating the pre-synthesised organic compounds into their structure. The decomposition processes are accompanied by release of nutrient elements into the surrounding environment, and the nutrients then become available to support new plant growth.

Biomass also serves as a source of nutrition for larger organisms such as herbivorous animals including humans (called consumers) that consume plant material as food. Some of the water, organic matter and nutrients that are taken in as food are released as waste, which again becomes food for new populations of microorganisms and plants.

The complex and interrelated sequences, consisting of production, consumption and decomposition of biomass, are a common feature of every ecosystem (see Figure 1.5), but the unique characteristics of any particular situation depend on the environmental setting. The physical geology and climate determine to a large extent what microbial, plant and animal species will thrive, and so we find around the globe recognisable natural systems of many types: prairie grasslands, tropical rainforests, deserts, boreal forests, tundra, to name a few.

The cyclical flows of energy, water and nutrients—from producers to decomposers and consumers, and back again—should be evident from this description. Along with the assured supply of energy, water and nutrients, it is this self-contained feature of ecosystems that largely contributes to their stability.

Even the most stable of ecosystems is not totally self-contained, not a totally closed system living unto itself. External components are also required, and inputs from outside the ecosystem are used to build up

**Figure 1.5**
An ecosystem is a defined area on Earth, subject to inputs from external sources, and outputs into other parts of the environment, but also self-contained in that a number of processes cycle energy, water and nutrients within the system. The result is a relatively stable, interconnected set of species and non-biological components

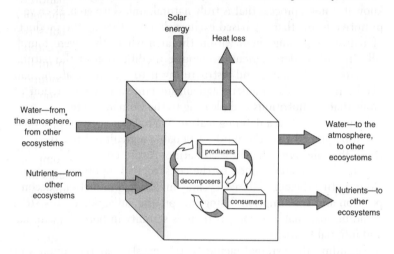

and maintain the internal cycling. Inputs do not however accumulate indefinitely, as some are lost or exported, moving through the original ecosystem into adjacent areas. Some solar energy is reflected back into the atmosphere while some is first converted into heat, which is partially radiated into space. Even the energy that is captured within the ecosystem (through photosynthesis to produce biomass) does not completely remain there. A part is lost through the release of heat during decomposition and some is exported (or imported) via wind, water or animal movement. Nutrients and water, too, do not remain exclusively within a clearly outlined ecosystem box. They can also move from one area to another, through wind, surface and groundwater flow and animal mobility (see Figure 1.5).

Nonetheless, a natural ecosystem can sustain itself over extended periods of time. It is sustainable both because the external processes provide a continuing source of needs for life that is characteristic of that system, and because the internal processes support a collection of life forms that coexist and depend on one another. Sustainability is one of the defining features of a natural ecosystem.[6]

## Agricultural ecosystems

Given this description, we can legitimately ask whether it is possible to think in terms of agricultural ecosystems. Clearly, agriculture as we know it is not a process that is truly natural, unless we go back to very primitive forms that are based exclusively on gathering the products of nature and using them within the area where they were found. All types of modern agriculture require modification of the natural ecosystem, redesigning and restructuring it to accommodate human needs. In particular, modern agriculture emphasises production of crops that are nutritious and pleasing to the human palate.

If sustainability is a defining feature of a natural ecosystem, then it could be argued that the most sustainable agricultural practices are those that are most aligned with nature. Again, consider Figure 1.5. Sustainability is maintained by the constancy of inputs and outputs along with efficient internal cycling involving production, decomposition and consumption. Using this picture, an agricultural system could be sustainable as long as there is stability in both the external and internal factors.

Regarding the external factors, this suggests that the supplies of energy, water and nutrients must be maintained at a level appropriate for efficient production of the crops in question. Solar energy is a 'given', but other forms of energy also contribute to raising a crop. Depending on the situation, the additional requirements could include human and animal energy, energy embodied in materials such as fertilisers and machinery, and energy in fuel. Likewise, water could in some ways be considered a 'given', at least in rainfed agricultural situations. However, for many farmers rainwater is a much less reliable input than is solar radiation. Where irrigation is available and used, its reliability is also an issue. Nutrients are always available internally, from the soil as part of the original mineral components or from decomposition of biomass residues from plants or animals. Nevertheless, the supply of nutrients may be considered insufficient, so additional amounts are frequently supplied from outside the system by using manufactured sources of inorganic fertilisers.

An important feature of a sustainable agricultural system then is that the supplies of inputs should be readily and sustainably available. This implies that there should be minimal dependence on the external, non-renewable sources and maximal utilisation of the internally available and renewable resources. Nevertheless, it is evident that the purpose of

#### Figure 1.6
The distinction between natural and artificial ecosystems is that a purely natural one is maintained by locally-available nature-provided inputs and is maintained in a state that is compatible with its surroundings. An artificial ecosystem requires an augmented supply of inputs beyond those from nature, and is maintained in a state that differs substantially from the equilibrium state that is compatible with the local environment

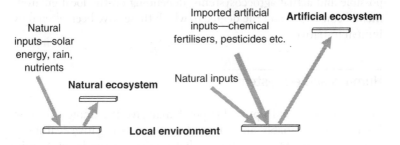

farming is to produce food and other products. As such, these products will almost always be 'exported' from the site. If plant and animal products are exported without internal sources of replenishment, the embodied nutrients are no longer available within the system to support new plant growth. This break in the natural cycling processes will then require external support, perhaps by demanding an additional input of nutrients from outside the system (see Figure 1.5). Nitrogen fixation is one means by which this essential nutrient can be imported naturally through organisms that convert atmospheric nitrogen into forms available to plants. At the same time, advantage can be taken of internal cycling of nutrients, by leaving crop residues in the system, allowing decomposers to feed on them and incorporating the biomass and nutrients into the soil.

Can we then legitimately think of agricultural systems as true ecosystems? Using our description outlined above, the answer has to be yes. Any agricultural system shares the dual nature of existing on imports and exports as well as on internal cycling processes. Clearly, however, no modern agricultural ecosystem is a natural ecosystem. Agroecosystems inevitably involve modification of nature by human design. In many cases, the greater the degree of human modification, the greater the requirement for external sources of energy, water and nutrients in order to maintain the altered state. We then see such input-dependent systems as artificially maintained in a state far from nature's equilibrium (see Figure 1.6). To do this brings into question whether such efforts are sustainable even in the short term but especially over extended periods of time.

To sum up, we will use the term 'agroecosystem' to describe a type of artificial ecosystem consisting of an area of agricultural land where there are common features in crop selection, and where broadly shared agricultural practices are employed. There are a whole range of possible and actual agroecosystems, depending on the local environmental conditions and the degree to which these have been altered by human initiatives.

## Humans and ecosystems

What has been said up to this point may give the impression that humans are invariably an unimportant or even negative feature in terms of sustainable ecosystems. That impression can be quite misleading. Several issues concerning humans and ecosystem are of great importance.

- Considering the wide range of ecosystems, there are some that involve wild nature in uninhabited parts of the world—wild nature without a human presence.
- In other ecosystems, the vast majority in fact, humans are an essential component, coexisting with the other various life forms. Agroecosystems are a prime example.
- It is precisely the presence and importance of humans that requires us to consider social and economic issues along with environmental issues when we are assessing the sustainability of agroecosystems.
- Human inputs in agroecosystems need not be a negative feature. If sustainability means carrying out agriculture in a way that honours the surroundings and operates within the constraints of local environmental conditions, then planning and action that respects these ideas can be directed toward sustainability.

The conclusion to this discussion then is that concepts of ecology applied to agriculture—what we can call agroecology—are essential for developing an understanding of sustainable agriculture. Furthermore, they are essential in planning methods of assessment or measurement to determine the degree to which particular agricultural systems are operating in a sustainable manner. We will refer back to these issues frequently in the subsequent chapters of this book.

## Notes

1. Throughout this book we will use the term low-income countries as an accurate but neutral term to describe countries that others may term 'the Third World', 'developing countries' or 'the South'.
2. The fundamental SI unit of energy is the joule (abbreviation J) and is used to describe energy content of fuels, energy used by humans and animals, and energy embodied in food and all other materials. Of particular concern in the present age is the consumption of non-renewable sources of energy such as coal, oil, and natural gas.
3. Herman Daly and other economists are at the forefront of developments of ecological economics, a discipline that attempts to incorporate environmental and social externalities into the economic system. See, for example, Daly and Cobb (1990) and Daly (1997).
4. Calorie is a unit of energy. The term 'food calorie' (usually just called 'calorie' (cal)) as used above is actually a misnomer, and one food calorie is equivalent to a kilocalorie (kcal) in energy terms. Another unit of energy, the one defined in the SI system, is the joule (J). The relation between the calorie and the joule is 1.0 cal = 4.18 J. Twenty seven hundred food calories is therefore 2,700,000 'true' calories, which equals $1.1 \times 10^7$ J. In a later section of the book, we will expand on the use of energy units.
5. The symbol $\{CH_2O\}$ is used as a simplified method of representing the complex organic matter macromolecules that are the components of biomass. In plants, much of this matter is carbohydrate, like cellulose, in which the ratio of carbon, hydrogen and oxygen atoms is 1:2:1 as shown in the formula.
6. Although we talk of a natural ecosystem as being stable, it is still subject to external events that may change its nature catastrophically—events such as fires or damage by flooding and ice. Over time the disturbed system may then return to equilibrium, one that may resemble the original system, but sometimes one that is completely different.

# 2 Sustainability Indicators

## 2.1 Purpose and Properties of Indicators

Due to its complex nature the assessment of any activity, as to whether it is being carried out in a sustainable manner, requires a detailed, multidimensional investigation, followed by a thorough description of the findings. Ultimately, the appraisal requires that the processes be followed over an extended period of time—it is the time element that distinguishes sustainability assessments from evaluations of many other activities. To be assured that certain agricultural practices are sustainable might require an investigation within a controlled system that extends over centuries. During the process of this extended study, one would follow all the relevant aspects of the agricultural situation—crop production, economic stability, the social situation—and at the end one could make a decision about whether the practices are sustainable. Clearly, this ideal experiment will not serve the needs of the present. For the sake of all humans on the Earth and for the Earth itself, we need to know *now* whether practices being followed *will be* sustainable. To satisfy this need, it is necessary to devise methods of predicting sustainability.

A careful selection and application of indicators of sustainability can be a first step in developing the essential comprehensive picture. As the word suggests, an indicator is a number or other descriptor that is representative of a set of conditions, and indicates or points to aspects of an issue. The measurement of body temperature as an indicator of human health is a good example to illustrate the ways in which indicators are used. A human body temperature substantially higher than 37°C usually suggests a health problem. While that single measurement is not sufficient to provide evidence of all types of illness, elevated body temperature is a simple indicator that a problem does exist for a wide range of health issues. Individual temperature measurement is further limited in that it does not provide any details about the nature of the health problem, but merely points to the need for further investigation, an investigation that might have additional qualitative and quantitative components. In the end, as a result of the detailed study, a comprehensive description of the problem and possibly some solutions can be produced.

> **Box 2.1**
> **Some well-known indicators**
>
> - **Body temperature**: an indicator of general well-being of humans and other mammals, measured in units such as degrees Celsius
> - **Gross Domestic Product**: a measure of the vitality of a nation's economy, commonly measured in United States dollars (US $) per capita per year
> - **Fuel economy**: a measure of the efficiency of a vehicular engine, commonly measured in litres per 100 km
> - **Crop yield**: a measure of the productivity of a particular crop, typically measured in tonnes per hectare

Likewise, appropriately chosen indicators can be a means by which sustainability is measured. As with body temperature, while measuring the status of a particular subject any individual indicator will be unable to provide a sufficiently detailed description of the complex and interrelated issues involved. Yet the simple individual value can act as a warning sign that leads to further study and action.
Indicators are useful in many ways:

- **As a management tool:** Indicators can be used to help improve the quality of operations that are being carried out. They can be effective in helping to identify processes or areas under stress, define opportunities for improvement, set priorities, allocate resources and provide a means of measuring accountability.

- **As a research tool:** The integrative nature of indicators makes them appropriate as a means of summarising complex data. The opportunity to compare quantitative data pertaining to different regions and over time becomes the basis for the development of hypotheses that describe factors contributing toward or against sustainability.
- **For educational and motivational purposes:** When an indicator integrates and simplifies the various aspects of a particular issue, it becomes a means by which information can be provided to stakeholders and the general public in a simplified and highly visible form. In some cases, the indicator value may be all that the public gets to know, or even needs to know; in other cases, the value may provide the rationale for further investigations. Publicising appropriate indicators can be a stimulus for action. For example, American law requires that industrial plants emitting toxic air pollutants report the specific contaminants being released. When such information became publicly available, environmental groups combine the data into an indicator showing total release of toxics by various companies. The media publicise the indicator information, using headings like 'The Ten Top Polluters'. The widely disseminated publicity generates a rapid response from companies on the list and as a result, emissions of toxic air pollutants are frequently reduced substantially in the following years.
- **For project assessment**: By setting objectives for a project, establishing indicators that can be used to measure attainment of the objectives and then making quantitative baseline measurements followed by corresponding measurements throughout the project, it becomes possible to develop an accurate picture of its progress and success.
- **As a planning and policy instrument:** Arising from the previous four purposes of establishing and using indicators is the possibility of using data obtained from particular studies for future planning and policy development. The policy aspects are closely related to the educational functions of indicators. If there are simple measures of the status of sustainable development in a given situation that have been established in a participatory fashion and are widely disseminated and publicised, policy makers can use these both to create and to justify policy decisions.

## One indicator or many?

There are various arguments in favour of having one or many indicators, which are summarised here:

- **One combined indicator:** In one sense, the ideal sustainability indicator would take the form of a single number that could be widely disseminated, and which would become visible to and understood by as many members of the relevant public as possible. Then, like a thermometer measuring temperature, multiple additional measurements could be made in different places at different times, various situations compared, and progress or regress followed as the years pass. Such an indicator would have to be a composite, incorporating data describing a wide variety of disparate issues. As such, it could sometimes be strongly influenced by a small number of factors (or even one single factor), and could therefore be misleading in describing the true situation. Alternatively, the average values for most of the factors could mask out one or a small number of different values that might be pointing to potentially serious problems. In either case, the message being conveyed by the single indicator is at best unclear and sometimes undecipherable or misleading.
- **A set of many indicators:** Using a different approach, one might wish to define a large number of indicators, each directed toward a specific, well-defined issue. In this case, the full range of 'good' and 'bad' values becomes clearly evident, but it is at the cost of increased complexity and reduced clarity. Many people would not be inclined to study the complete data set, and even when they did, there would be difficulties in interpreting, integrating and using the various pieces of apparently unconnected information.
- **An intermediate approach:** An approach to using indicators that attempts to reconcile the advantages and disadvantages of single and multiple indicator methodologies involves beginning with careful collection of as much detailed data as possible. These raw data are compiled in a way that should make them widely available to those who would wish to examine the subsequent processes of data treatment. Based on the primary information,

indicators are then created using methods that will be described in a later section of this book. The indicators are evaluated and key ones selected to be combined in a rational way, forming a set of composite indicators. Further aggregation of the composites may be used to generate a smaller number (even one, if desired) of broad-based indicators. If the entire data set is made available along with a transparent description of the procedures used for generating and aggregating the indicators, then the information can be employed at various levels, depending on the purpose for which it may be required.

Details of the processes of generating and aggregating indicators are described later in this book.

## Types of indicators

Indicators are used to display different types of information, and can be grouped in various categories. Frequently, three categories are defined:

- **Pressure (or stress) indicators** measure activities or processes that have the potential to drive the situation under study towards a particular conclusion. For example, if one is attempting to measure groundwater availability for agricultural purposes in a given area, a pressure indicator might be the total annual amount extracted in that defined region, measured in cubic metres per year. A series of such measurements made over an extended period of time might show increasing demand that could eventually lead to depletion of the groundwater resource. In this case, the water-use parameter points to a possible cause of stress that could adversely affect a given system.
- **State (or composition) indicators** are indicators that measure the state of the system at a given time. Using our example regarding groundwater availability, measurements of the depth of the water table would be a state indicator. While information about the amount of water extracted each year indicates a potential problem, the state indicator is a more direct indication as to whether the rate of extraction actually exceeds the rate of recharge. While it provides a direct measure of the current

situation, a state indicator does not provide information as to the cause of a particular problem.
- **Response indicators** are the third category, and these show what is occurring to counteract the pressure—usually some type of remedial measures that are designed to alleviate an adverse stress. Therefore, with water availability for irrigation, a typical response would be to initiate various types of water harvesting or water efficiency measures. If these are quantified in some way, the data can be transformed into a response indicator. In many cases, response indicators provide positive information in reaction to indications of stress or of an unsatisfactory state.

---

**Box 2.2**
**Types of indicators**

**Pressure indicators** show potential or actual stress that could lead to problems.
**State indicators** measure conditions within the system being studied at a given point of time.
**Response indictors** describe measures that are being used to alleviate the stress or to improve the adverse state within a given situation

---

In a sense, we can think of the three types of indicators as measures of the past, present and future. Pressure indicators show what has been done, and continues to be done, to generate a particular situation. Carefully chosen state indicators describe the present status. Response indicators show what is being done to relieve a stress, and therefore look toward future improvement.

In any comprehensive study, including studies of sustainability, it is useful to include indicators in all three categories. Together these can provide a more comprehensive picture of the history, current state and future prospects of the situation or process under study. Taken as a group, a comprehensive set of indicators involving all three types is frequently referred to as a set of Pressure-State-Response (PSR) indicators.

This division of indicators into three categories is widely but not universally used. Among other proposed classifications of indicator types is that used by the European Environment Agency. This agency assumes a five-fold framework (DPSIR) that considers cause-effect relationships between the interacting components of economic, social and environmental systems (Smeets and Weterings 1999). Using an example of indicators to provide information about drinking water quality, the five divisions are:

- Driving forces of change (e.g., industrial production statistics)
- Pressures on the system (e.g., discharges of wastewater)
- State of the system (e.g., water quality in lakes and rivers)
- Impacts on population, economy and ecosystems (e.g., water unsuitable for drinking)
- Response to the pressure (e.g., watershed protection/environmental laws)

Using this classification, the relation between the various types of indicators becomes clear. Each component of the sustainability cycle influences the next one. In particular, the initial issue described as the driving force for a sequence of consequential effects can eventually lead to a response that will in turn have an effect on the initial force. In the case of indicators related to drinking water quality, the response of watershed protection should bring about positive changes as the modified cycle revolves again. The relations and feedback aspects of the five DPSIR components are illustrated in Figure 2.1.

This DPSIR classification is really a modified version or an expansion of the PSR system, where pressure and state indicators have been subdivided into somewhat more specific components.

## Secondary indicators

In some cases it is difficult or impossible to make a direct measure of a particular component of sustainability. For example, we will see later that the most direct measurement of stability of agricultural production is a time series showing yield over an extended period of time. Time is the issue, and one would not normally expect to measure a systematic trend showing changes in productivity unless data were available covering several decades. In most cases, there are only limited data available from the past, and collecting data well into the future would necessitate a lengthy process until sufficient information is obtained. In cases like this, one could look for a surrogate indicator

**Figure 2.1**
The components of the DPSIR system of classifying indicators. The example used for each component is that of drinking water quality

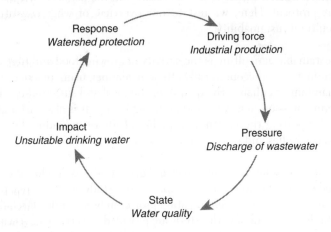

that could be used to predict possible future changes. One type of surrogate indicator that can be used to predict stability of crop production involves assessment of soil properties—depth of topsoil, organic matter content and nutritional status. Where these properties are less than ideal, one could expect that maintenance of high yields would be unlikely, unless, of course, remedial action is taken.

Another type of secondary indicator, sometimes called a crypto-indicator, is one that on the surface has no connection with the process at hand. However, such indicators may be used to reveal an otherwise hidden issue. Increased attendance at a temple, mosque or church has been suggested as a way of indicating a real or perceived problem in a local society—sometimes a problem that might otherwise go unnoticed.

## 2.2 Steps in Developing Indicators

### Conceptual overview and category selection

It is important to re-emphasise that, in any programme of assessment, indicators should not be chosen without first developing and clearly expressing a conceptual overview of the situation and the objectives of the measurement process. This is essential as a basis for subsequent decisions when it comes down to working out the details of the assessment protocol. Here, we restate our overview of what constitutes agricultural sustainability:

> Sustainable agriculture is the activity of growing food and fibre in a productive and economically efficient manner, using practices that maintain or enhance the quality of the local and surrounding environment—soil, water, air and all living things. It is also sustainable in supporting the health and quality of life of individual farmers, their families and the community as a whole.

This defined vision is clearly broad and general. It was developed over a period of several years during a comprehensive study of agricultural practices in and around a major irrigation project in South India. (Details of the findings surrounding this study are provided in Section 4.7.) Having put out a conceptual statement, a second step is to further refine and expand the overview to set out more specific areas that require investigation. We call these areas categories. Appropriate indicators chosen within these categories then work together to provide a description of the total situation.

Recognising that agriculture is a process of food and fibre production as well as a way of life, the categories are chosen so as to reflect its various dimensions.

- **Productivity**  For the needs of the farm family as well as to satisfy global food requirements, any sustainable agricultural system must be capable of producing high yields.
- **Stability**  It is necessary that the high level of productivity be maintained over an indefinite period of time. This requires that the quality of the resources on which production is based also be maintained and even enhanced.
- **Efficiency**  To be sustainable, all the resources required for agriculture—human, animal and material—should be used in a way that is not wasteful, but maximises output per unit input. This is especially true of non-renewable resources.

- Durability       Any crop production process is from time to time subject to stresses of various types, such as those due to water or to pests. Sustainable systems are intrinsically resilient in the face of such stresses.
- Compatibility    Sustainable agriculture should fit in with the human, social and natural environments where it is located, maintaining and enhancing the health of these environments.
- Equity           Agriculture should promote a good quality of life among the various individuals involved in the farming activities and within families. This includes having consideration for the standard of living, health and education of all people in the community.

The overview and categories suggested here are meant to provide a comprehensive assessment of agricultural sustainability. Other workers in this field have developed different, but usually similar, frameworks that can be used as scaffolding for the construction of indicators. Building on a framework of this type, we can then set out to develop individual indicators that can point to issues of sustainability.

## 2.3 Desirable Properties of Individual Indicators

Having set out categories to be assessed, indicators are chosen, modified or often developed afresh, making selections from the three PSR types as described above. The careful choice of indicators is critical in order to achieve the desired goals of the investigation. In this process, certain desirable properties are kept in mind:

### Relevance, quality and reliability of indicator data

Having defined the situation that is to be studied, and the overview of the study, a central criterion to be used in developing indicators is to ensure that they are directly relevant to the objectives of the study. Because an overarching objective, such as achieving sustainability, may be very general and even vague, it is essential that the sub-categories within the general objective be more specific and clearly defined. With this starting point, it becomes possible to choose and develop appropriate relevant indicators. For example, in a study on soil quality in a particular agroecological situation, the question must first be asked regarding what the quality issues are. These may relate to topsoil loss by erosion, to organic matter content, to soil-based nutrient supply, or to contamination by residual pesticides or other organic or inorganic contaminants. Whichever of these categories (or other categories) is identified as being central to the study, one can then make a rational selection of indicators that will provide information about the category(s).

Quality and reliability are also requirements. The availability of plant nutrients, including the major nutrients, nitrogen, phosphorus and potassium, is a case in point. A number of different soil tests have been developed for each of these and, depending on the soil type, crops being grown and other environmental factors, particular tests are a reliable indicator of nutrient availability. Other tests that measure concentrations of the same nutrient in different ways may provide misleading information in particular situations.

In obtaining data, whether one is using physical or chemical measurements, there are errors associated with the results obtained. Errors are an intrinsic feature of every measurement process and arise from sampling as well as from the analytical steps themselves. Knowledge of these uncertainties is important in order to determine whether apparent differences obtained at various sites or at different times are truly significant. Likewise, with economic or social data collected by surveys, there are always errors associated with the selection of the sample of individuals in the survey. The larger the sample, the more reliable the data, but one can always make a statistical assessment of the margin of error depending on the sample chosen. Even more serious than the 'normal error', which can be estimated using statistical methodology, is the possibility of bias. Bias is not eliminated simply by choosing a larger sample, but only through careful planning to ensure that all components of the population under study are systematically provided an opportunity to be part of the sample.

## Availability and ease of collection of indicator data

In many agricultural situations it is possible to collect a vast amount of data connected with issues surrounding food production. In fact, sometimes these data are already available, having been collected previously by other workers in the area. There is a temptation to use all the readily available data, even though they may not be the most appropriate with respect to the objectives of the study.

International bodies, national governments, and local non-governmental organisations are all excellent sources of data that can be used in studying environmental, economic and social issues related to sustainability. It is only in the last two decades or so, however, that sustainability has been studied as a specific subject. Therefore, while we may have considerable information about economic activity in particular spheres, there is often no indication as to what component of this activity relates to processes that can be considered to be sustainable. Likewise, in earlier years agricultural statistics focused on issues like crop production, amount of land being brought under cultivation, fertiliser consumption, etc., implying that growth in any of these areas was a sign of progress. It is frequently found that the data required to measure sustainability have never been collected, and there may be no design or protocol yet determined for their collection. Adding to this difficulty is the fact that some sustainability-related information is subtle and difficult to quantify. For these reasons, considerable effort must go into developing and collecting new sources of information, taking into account practical issues like availability and ease of collection.

Recent technological developments have opened up possibilities for obtaining new types of data. Here too, care is required to ensure that the information is meaningful and relevant to sustainability issues. For example, in studying the role of uncultivated land (including land remaining in a natural forested state) within an agricultural area, it is tempting to make use of readily available information obtained by remote sensing. The proportion of forest or other uncultivated land in a certain region can be determined simply and relatively inexpensively without having to resort to expensive and time-consuming field investigations. Remote sensing data are useful in many situations, but in some cases the information may be inadquate and sometimes misleading. For example, most data obtained from satellites do not tell the age, composition or condition of the forest, all of which may be essential features required for considering it as a source of biodiversity or as a habitat for other plant and animal species.

## Stakeholder participation in development

As pointed out above, depending on what is being measured, some of the information needed for creating an indicator is available from a variety of sources. Government statistics can often provide the information required, especially for macro-indicators based on data related to items like population, education, health, agricultural production, and average income. Supplementary information of this type is also available from supranational organisations such as UN bodies like the Food and Agriculture Organisation (FAO), World Health Organisation (WHO), etc. or the OECD (Organisation for Economic Cooperation and Development). Non-governmental organisations can provide additional data, sometimes at the macro level but often related to more local situations. It is important to recognise that there can be a bias in the statistics obtained from any of these sources, in the selection of information provided and in the way it is presented. This is one of the reasons why it is essential at every stage that sources and methodology used for creating indicators should be available.

Obtaining information in the field is another matter, and the entire operation should involve participation of the local population. The process of developing micro-indicators used to measure agricultural sustainability in a local area begins with the collection of field-based basic information, before one can begin the processes of converting data into quantitative information. At this initial data collection stage, it is important that 'experts' and the local population work together. The experts should bring in background ideas, often technical and specialised in nature. They should also have a sense of the context of the assessment process—what related information has been collected elsewhere, what are some of the problems and pitfalls likely to be encountered, and how these can be dealt with. The experts will have the technical knowledge needed to carry out the required manipulations of the data, creating indicators and doing scaling and aggregation.

On the other hand, the local people have an unrivalled sense of the nature of their own surroundings and of opinions and worldviews shared by the population. They also bring a type of realism to the process, in identifying some of the issues that might be addressed, and in making decisions about what is possible as a response to problems and what cannot be done.

Evolving from extended open discussions and the back-and-forth of listening and presenting ideas by both groups, the resulting combined contribution can make an assessment much more valid and meaningful. Even more than that, a participatory approach ensures that the local population is kept aware of and shares a common view of what the assessment is about. This enhances the possibility that they will use the results towards taking the positive steps that may arise from the recommendations of the study.

Once again, the possibility of bias adversely affecting the assessment must consistently be kept in mind and avoided. Bias may arise from the expert side, usually beginning with the conceptual vision that forms the basis of the study. In a way, such bias is inevitable and the 'solution' is simply to clearly state the worldview at the outset. Bias from local participants also occurs in many studies. For this reason, it is important that those who are involved come from a diverse and representative group. Care must be taken that the study does not reflect the views of only of an elite component of the community, or of only men or women, or of a particular age group. The need to be sensitive to gender issues and concerns is frequently not fully considered and is of particular importance in almost every agricultural situation (see Box 2.3).

---

**Box 2.3**
**A guide to planning gender-sensitive indicators**\*

1. Ensure that there is a *people focus* as well as a technical/environmental component to the objectives of the study and to all assessments made. In addition, ensure that the people focus differentiates between men and women. This will require that careful attention be paid to the differences in the roles, responsibilities and resources that affect both the participation and/or the benefits for women and men.
2. Ensure that the description field for each major output refers to both women and men and the gender inequalities that are to be addressed. Do women and men have different responsibilities?
3. Ensure that the description field refers to the way in which the planned activities will address the different needs and priorities, including women's access (or lack of access) to the resources necessary for their participation and benefit.

*(Continued)*

*(Continued)*

4. Ensure that both women and men are involved in the planning activities, developing of indicators and collecting data. As it may sometimes be more difficult to reach women in rural areas, it may be important to design particular ways of reaching women.
5. Ensure that the immediate impact or benefits as well as the longer-term outcomes for both women and men are included in the discussion of effects. Ensure that any prior assumption of gender-neutrality is valid when assessing the outcomes of the project. Determine whether possible unplanned effects and outcomes that might be negative for either women or men have been anticipated and addressed.
6. Ensure that both women and men as well as organisations with a gender mandate are included in the early stages of planning so that a balanced set of issues is addressed. The stakeholders should include local persons of both genders, as well as national and international NGOs that deal with gender and women's issues.
7. Identify and use quantitative and/or qualitative indicators to measure the gender sensitivity of all aspects of the project. Gender-sensitive indicators provide the clearest evidence that gender roles and responsibilities, and particularly the needs of women, have been carefully considered and addressed.
8. Identify relevant quantitative and qualitative indicators to measure the participation of women and men at every stage of the planned sustainability study.
9. Include a longer time frame in the measurement of output. Identify indicators that can measure the outcome after a three to five year period.
10. Ensure that there are appropriate plans, including budgetary arrangements, to allow for the gender disaggregation of data at all levels of the study.

\* Paraphrased from a guide developed for the Food and Agriculture Organisation of the United Nations (Kettel 2001).

# Sustainability Indicators 71

*The Community Indicators Handbook* documents a process of ten steps (as outlined below) for a participatory approach to developing a broad-based indicator set (Meadows 1998). Various features of the approach are appropriate in diverse studies around the world.

1. *Select a small working group responsible for the success of the entire venture*: The working group needs to be multi-disciplinary, with strong ties to the community or audience for whom the indicators are intended. The working group is most effective when it combines experts and non-experts from the outset, but the critical element is long-term commitment to the process.
2. *Clarify the purpose of the indicator set*: Make clear whether it is meant to educate the public, to provide background for key policy decisions or to evaluate the success of an initiative or plan. Different purposes give rise to different indicators and publication strategies.
3. *Identify the community shared values and vision*: The indicator set must be able to speak to the hopes and aspirations of the people it is meant to serve.
4. *Review existing models, indicators, and data*: The working group takes a look at other indicator projects as examples to learn from. It also reviews what indicators are already published locally and what data are generally available.
5. *Draft a set of proposed indicators*: The working group draws on its own knowledge, the examples it has collected, and the advice of outside experts (if needed) to prepare a first draft. The draft may go through several revisions before it is ready for the next step. In particular, initial indicator sets tend to be very long. In later drafts, they need to be pruned down and made more focused and practicable.
6. *Convene a participatory selection process*: The draft indicators need to be presented to a broad cross-section of the community for their input. This process serves several important goals. It educates the participants, gathers their collective creativity and expertise, and makes them stakeholders in the success of the project. Often it also gives rise to new relationships and alliances among the participants and can even generate new action initiatives to address problems identified by the indicators.

7. *Perform a technical review*: An interdisciplinary team of knowledgeable people sorts through the proposed final draft indicators and selects from among them for measurability, statistical and systemic relevance, etc., trying to stay true to the intentions and preferences expressed by the citizen review process. The technical review helps to fill in gaps, weed out technical problems, and produce a final indicator set that is ready to be fleshed out with data.
8. *Research the data*: At this stage, the indicators are usually subject to additional revision, driven by data concerns and new learning.
9. *Publish and promote the indicators*: This requires translating them into striking graphics, clear language, and an effective outreach campaign. It helps to link the indicators to the policies and driving forces that affect them, to illustrate their linkages, and to point to the actions that can be taken to improve them.
10. *Update the report regularly*: Indicators make little difference, or indeed little sense, if they are not published periodically to show change over time. This requires an institutional base that can be relied upon to reproduce steps 8 and 9 on a regular basis, and go back and revisit the other steps as needed. Each new version of an indicator report becomes an opportunity to revise the indicators, develop new research methods, and add linkages. If performance targets have been set, they can be assessed and, if necessary, adjusted. And when targets are met, celebrations can occur!

These steps may sound daunting, but they are being put into practice by hundreds of community- and regional-level indicator movements around the world.

## Convertibility of indicator data into quantitative terms

Certain types of information are intrinsically quantitative, but in order to use this information it may be necessary to convert it into other forms that will be more understandable to the public. Using a common scale is one useful way of making it possible to bring together several types of information. Simple scales such as percentage or a zero-to-ten scale are sometimes used. In other cases, it can be helpful and attractive to use values that are based on physical or economic realities, like areas of land or monetary units. There are many other possibilities.

The following are examples of units that are usually familiar to at least some members of the population. It will be obvious that these units can be used to describe some environmental and/or economic issues. They will not be useful for indicators functioning in the social realm.

### Monetary units

Expressing economic output in United States dollars (US $) is a common practice, widely accepted by policy makers around the world. It may not, however, carry much resonance with the people of some countries where American currency is an unknown quantity.

### Resource depletion time

An early and widely publicised environmental study specifically focused on issues of resource depletion (Club of Rome 1972). The Club of Rome made use of the idea that there is a limit to the amount of non-renewable resources (which we have termed 'natural capital') available in the world. The Club published and has continued to publish extensive data showing these limitations in terms of years of resource remaining. To do this, trends regarding levels of population and consumption were predicted. An example of this kind of thinking is shown in Figure 2.2.

This figure, presented in the 1991 publication, shows predictions of the global supply of arable land available for agriculture, as well as the amount of land required to feed the world's populations, assuming certain levels of productivity. A simple conclusion from the figure is evident by taking the point at which the line showing productivity (at 1990 levels) intersects the line showing supply of arable land. These two lines cross at approximately the year 2000, indicating that the world would go into a food deficit situation at about that time. The plot also shows that the food deficit could be postponed to about 2030 or 2060 if productivity doubled or quadrupled. Clearly, this approach has the potential for much more dramatic public impact than would be the case with a detailed discussion of the raw data leading to the stated conclusion. Note, however, that this prediction made by the Club of Rome has not been borne out in reality. This once again

## Figure 2.2
A graphical prediction of the amount of agricultural land (in billion hectares) required to satisfy global food requirements at current and accelerated levels of productivity

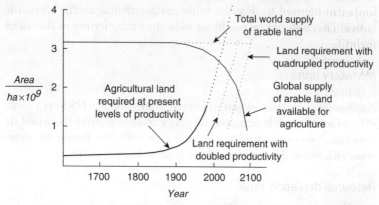

Source: Redrawn from Club of Rome (1972), p. 50.

emphasises the importance of ensuring the validity of the data and the predictive methods if one is not to overstate (or understate) a problem. Unjustified 'crying wolf' can only generate cynicism and potential future inaction when other such stark indicators are presented.

### Energy units

Energy is another parameter that can be used for expressing values of many physical activities. It can be measured using a variety of units, some quite graphical and visible such as tonnes of coal or barrels of oil. As noted earlier, energy can also be expressed using the fundamental SI unit called the joule (J). Energy is an especially useful concept for dealing with natural and manufactured capital, as well as with the agricultural processes that make use of such resources. In a later section, where we examine how to measure the efficiency of agricultural processes, we will discuss energy issues in more detail.

### Land area

Building on the idea that energy consumption is a central feature of all physical activity, Wackernagel and Rees (1996) have developed the concept of the *ecological footprint*. The ecological footprint value is a measure of the area of land required to sustain the life of the average person living within a particular setting. The single aggregated ecological footprint index accounts for resource consumption and waste assimilation in terms of productive land needed to maintain life-supporting activities at a specific level. Calculations can be done for individual persons by taking into account all the physical activities of their life: type of shelter, diet, travel, employment situation and general consumption habits.

More commonly, calculation is based on the *average person* living in a defined area. Multiplying values for the average person by the number of people in a city, region or nation gives the footprint for that defined area. The total ecological footprint is therefore a function of population and per capita material consumption. The conversion to units of land area (hectares) begins with a calculation where material goods are converted to energy units, which are then converted into carbon dioxide release, assuming that the energy is generated by combustion of fossil fuels or biomass. The carbon dioxide release is finally given a spatial dimension by estimating the area of land needed to grow sufficient biomass to assimilate the carbon dioxide produced through the consumption processes.

### The ecological footprint

The calculated values of footprints for nations are estimates of the amount of biologically productive space (mostly land) needed to support all the human activities involved in production, consumption and waste disposal. Table 2.1 summarises information for selected countries, with the footprint calculated on a *per capita* basis.

On a single country basis, positive values of ecological surplus/deficit suggest that the country has sufficient regenerative potential to accommodate the consumption activities. This could be evidence of ecologically sound policies. Alternatively, some countries are well endowed with abundant ecological capacity and may therefore, operating independently, be able to sustain their current population at a relatively high level of resource throughput. On the other side, negative values are an indication that physical resource use is unsustainable. Where a country is running a deficit, it means that, on its own, it has insufficient ecological reserves to support the population with current patterns of consumption.

Using ecological footprint calculations in another way, it is estimated that, worldwide, the carrying capacity of the Earth is equivalent to

**Table 2.1**

Ecological footprint per capita in selected countries. The population is the 1997 value, and other information is based on 1993 statistics. The available capacity is the total regenerative capacity of land in the country divided by the population. Ecological surplus or deficit is the difference between available capacity and ecological footprint.

| Country | Population in millions | Footprint ha/capita | Available capacity ha/capita | Ecological surplus (+) or deficit (−) ha/capita |
|---|---|---|---|---|
| Bangladesh | 126 | 0.5 | 0.3 | −0.2 |
| Brazil | 167 | 3.1 | 6.7 | +3.6 |
| China | 1,247 | 1.2 | 0.8 | −0.4 |
| Colombia | 36 | 2.0 | 4.1 | +2.1 |
| France | 58 | 4.1 | 4.2 | +0.1 |
| Hungary | 10 | 3.1 | 2.1 | −1.0 |
| India | 970 | 0.8 | 0.5 | −0.3 |
| Italy | 57 | 4.2 | 1.3 | −2.9 |
| Japan | 126 | 4.3 | 0.9 | −3.4 |
| Nigeria | 118 | 1.5 | 0.6 | −0.9 |
| Russia | 146 | 6.0 | 3.7 | −2.3 |
| South Africa | 43 | 3.2 | 1.3 | −1.9 |
| UK | 59 | 5.2 | 1.7 | −3.5 |
| USA | 268 | 10.3 | 6.7 | −3.6 |

Source: Data from Wackernagel et al. (2002).

there being 1.7 ha of land available per person. On this scale, any nation or individual whose footprint is greater than this is putting the Earth into a global ecological deficit. It is clear from this abbreviated list that the average person in some (mostly wealthy) countries is consuming an excessive amount of the world's resources.

## Integrative capability of the indicator

To provide information corresponding to the three components of the sustainability tripod, many indicators may be required and many have in fact been proposed. The amount of data available to develop an indicator depends on the issue being studied, and on whether the study is at the macro or micro level. When the data are converted into indicators, some can be highly specific and focused on a particular closely delineated subject, while others bring separate aspects of an issue together in an integrative fashion.

Integrative indicators have at least one advantage in that the single measure covers a range of subject matter and may, on account of its simplicity, be more acceptable to policy makers and the general public alike.

There are two approaches used for developing an integrative indicator. One method involves collection of many pieces of individual data and combining these together in some clearly defined way in order to produce a single number that is an aggregate of all the data collected. Such composite indicators, usually termed indices, are widely used for reporting information obtained in environmental studies. The ecological footprint, for example, is a unique type of integrative index. Other examples include the various air quality indices (AQI) that have been constructed by combining detailed information about the atmospheric levels of sulphur dioxide, nitrogen oxides, ozone, volatile organic hydrocarbons, particulate matter and other parameters. An integrative AQI value is then useful for tracking general trends in atmospheric chemistry throughout a given year and over a range of years. It is useful to researchers and policy makers and is a source of information for the general public. For example, it can be used as a health advisory indicating the need for reduced physical activity on days when there is a high level of pollution, or it can be used in urban areas as justification for a call for remedial action such as restricting the use of motor vehicles on bad air quality days.

A number of economic and social indicators are also composites. The Human Development Index, designed after exhaustive study by the United Nations Development Programme, is one such indicator. It takes into account information on health, education and economic status. This indicator will be described in more detail later in this chapter.

Integrative indices essentially report a single piece of information, but that information takes on a particular value because of its relation to all the factors from which it has been derived. For example, the Gross National Product (GNP) of a country, calculated on a per capita basis, has many flaws but it does give a broad measure of all the varied economic activity occurring within that country in a given year. Some biological and physical indices serve an intrinsic integrative function. The concept of Biochemical Oxygen Demand (BOD) is based on the idea that degradable organic matter plus any other chemicals that can be oxidised will consume oxygen in a water body. A large BOD value occurs in highly polluted water with low oxygen content. The anoxic conditions are, in many situations, considered undesirable. A BOD analysis done in the laboratory measures the additive effect of all the possible contributions to this condition, without identifying any specific one.

In the biological sphere, declining populations or extinctions of particular species can be considered to be an integrative index that points to possible changes in several factors: habitat loss, pollution, introduction of predator species, etc. In fact, it is very likely that the reduction in numbers occurs due to a combination of these factors.

While the advantage of integrative indices is clear, it is not uncommon that some of the important specific pieces of information become buried and are lost in the single aggregate value. The important information can be uncovered only by examining the individual data values that are combined in calculating the combined index. This need not be a disadvantage, however, if the indicator is used for its true purpose, which is to point to the broad substance of the issue. The integrative indicator is a starting point that calls for a search to obtain further information.

### Sensitivity of the indicator to changes over time

Indicators are essentially developed for purposes of comparison. Most importantly, the comparisons to be made are with respect to an ideal—in the case of sustainability indicators the ideal being set by the overview or concept of what sustainability should mean. Therefore the indicator should be able to point to changes occurring as components of the situation move towards or away from the defined goal. It is therefore important that collections of data be made on a timely basis at frequent enough intervals, in order that an accurate picture of any direction of change can be observed.

In order to be able to distinguish between cyclical changes, the period of analysis must also continue over an extended period of time. Even robust national economies go through successive series of years that are relatively good, then relatively bad (boom and bust years), but what is important are the long-term trends. Agriculture too may experience cyclical ups and downs, due both to climate (several dry years, or an extended period of good rains) and to the economic situation itself. A 10-year period of increasing productivity may be due simply to the introduction of a new higher-yielding variety whose yield potential may for various reasons not hold out over time. A good indicator should be able to distinguish periodic or cyclical changes from the broad trends that show either positive or negative change in the long term.

The requirement to have measurements over an extended time frame is sometimes difficult to meet, especially when the duration of the study is limited by financial or other constraints, or when it is necessary to rely on previously collected data. There are, however, some long-term trends that can be reconstructed from geological and biological data, such as information about climate change gained from observing tree rings or from studying changes in the fossil flora and fauna in lake sediments (this paleolimnological information can provide an environmental picture that extends back hundreds or even thousands of years). In some jurisdictions, limited amounts of economic and social data have been kept in historical records over the past century or two. All this information can be used to construct so-called baseline values and a time series that enhances the value of measurements that are made in the present.

## Ability of the indicator to monitor changes across location and situation

In developing indicators for a particular situation, we may have some ideal attributes in mind. It is important to be clear regarding how the ideal attributes should be measured. Comparison of indicator values in different settings and with the ideal then requires that the same kinds of data be obtained in all the situations that are being compared. This calls for attempts to standardise the details of indicator design.

Soil quality data are a good example. As noted earlier, for every nutrient there are numerous analytical methods prescribed for analysis. Differences in the methods are found in the protocols for soil sampling, soil storage and pre-treatment processes, and the actual analysis procedures. The major nutrient, nitrogen, is a case in point. Opinions vary as to what fraction of soil nitrogen should be determined; whether it should be ammonium, nitrate, organic nitrogen or all of them; and whether the total amount needs to be determined or only that which is deemed to be readily available to growing plants. Clearly, comparisons between different areas require that there be agreement concerning details of appropriate methodologies.

The same requirement applies to all kinds of economic and social information as well. How do you measure literacy? What is the best measure of average educational level of a population? For these and many other issues, in order to create a quantitative indicator, there is need for a consensus about data collection and manipulation. The word 'manipulation' is appropriate here, as raw data has to be converted into a visible and understandable form using logical and mathematical steps (but with a subjective aspect as well). But the term manipulation also has a well-earned negative connotation. As with any situation where statistical information is used, it can be possible to manipulate the information in such a way as to create an impression that may not be justified. This negative aspect of manipulation has been misused in many instances around the world by researchers and policymakers alike. Wide-ranging discussions among experts working on a particular project, and with those doing similar work in other locations, help to establish consensus regarding appropriate protocols that minimise bias.

There is one more aspect regarding what is necessary in setting out the spatial dimensions of an indicator. That is the aspect of defining the scale of the space being studied. Here we use the word 'space' in a broad sense, to indicate the portion of a given area or population that is being studied. Where population well-being is measured, there could sometimes be a general improvement reflected in average income levels that masks increasing disparity between rich and poor. There may also be improvements in the level of education that hide deteriorating circumstances for some groups such as women or ethnic minorities. These are situations where working with a broad-based and truly representative group of stakeholders is required throughout all the stages of the study, in order to ensure accuracy and fairness. If sustainability is defined in terms of a system that provides adequately for a population, then the distribution as well as the overall amounts must be taken into account.

### Predictive ability of the indicator

Indicators that can be used to predict future consequences of continued action are highly valuable to policymakers and managers of all types, including to individual farmers. In many cases predictive characteristics are derived from the ability of the indicator to make comparisons over time and space. In the time-scale, creating a series of values that follows

**Figure 2.3**
Gross codfish catch (in thousand tonnes) in the Canadian Atlantic fishery located in Newfoundland. After the 1991 season (dashed line), the evidence of declining stocks was the impetus for increasing government action to impose strict quotas in the following years.

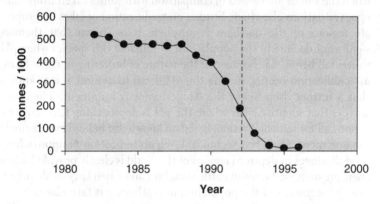

Source: The Department of Fisheries and Oceans, Government of Canada.

a systematic trend can enable extrapolation of the trend into the future. Emphasis must be placed here on having a clear and systematic trend, and on not extending the period of extrapolation too far into the future. In order to use the trend line, threshold values are often selected to indicate a need for some kind of action.

The Club of Rome plot (Figure 2.2), predicting the limitation of arable land supply as it affects the ability of the world to feed itself is an example of prediction using plotted time-based data. Another example of this type is found in the fishing industry. In Newfoundland, on the east coast of Canada, declining cod stocks over the years pointed to a catastrophic loss of the species due to overfishing (Figure 2.3). The loss has been so great that the population was on the verge of economic, if not actual, extinction, and there was a clear need for political action to restrict cod fishing in certain areas.

Note that the indicator 'gross fish catch per year' is an integrative one that is based on more than just the extent of fishing, and also includes potential information about other issues that determine catch, like loss of habitat and changes in water quality. In this particular instance, however, the major factor affecting the value of the indicator was the amount of fishing being done.

The use of a time series to develop a curve that shows trends into the future is perhaps the most obvious way of using indicators for predictive purposes. Individual values taken in the present can, however, sometimes serve the same purpose on their own. This is especially true when the values are viewed in comparison with values taken from analogous situations elsewhere. Studies on acidification of lakes in temperate regions of the northern hemisphere have shown that there is significant decline in the populations of valuable fish species when pH values fall below 4.5. Because of the nature of buffering in these lakes, as acidification occurs, driving the pH down to around 5, it is known that a further drop below the danger point is imminent. Therefore, surveys that identify lakes where the pH is approaching this sensitive region call for remedial action, based on knowledge behind even a single measurement. On a less technical level, an indication of approaching food shortages in deprived regions of the world is clearly provided when there are increased rates of cattle slaughter and when large-scale migrations of segments of the population into other areas take place.

### A clear relationship to vision, action and policy

Carrying out a study of sustainability can be an enjoyable intellectual exercise, but the real purpose has to go beyond this. Sustainability is always assessed for a purpose. And that brings us back to where we started—the need for a well-defined overview that includes the objectives of the study. After all, anyone who wishes to study sustainability will also want to promote sustainability. Indicators should therefore be action- and policy-oriented. All the desirable properties we have described here are directed to that purpose.

The goal is to identify and design indicators that are based on a shared vision, one that has a foundation of sound knowledge concerning sustainability issues and is at the same time closely connected with the practical experience of people 'on the ground'. The indicator must clearly describe the pressures, current status and responsive actions related to the area, project or process under study. Each indicator should point to an ideal value, and compare the current situation to that ideal. Every step in carrying out the study should be done openly and transparently, with a willingness to alter thinking, even to change direction as the study continues. In fact, iterative action is called for.

Using the best knowledge obtained by participatory means, a plan is devised, indicators are identified and data collection begins. While

#### Figure 2.4
The steps in developing and carrying out an indicator project. Note that the project has a starting point and a conclusion, but in the process there are several opportunities for consultation, feedback and reworking ideas

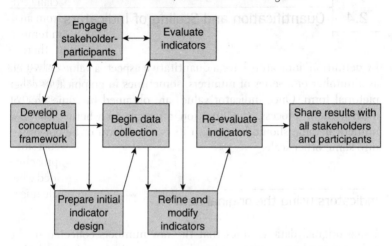

working on the assessment and afterwards, there will be many opportunities for rethinking and sometimes starting over. When an acceptable set of data (it will never be a final or a perfect set) has been obtained, the information is converted into indicators that can be clearly understood by the stakeholders. Because there has been broad participation from the outset, it is realistic to expect that action and new policy can emerge from the process. This will require careful attention to methods of dissemination, including meeting with people at all levels—farmers, extension workers, researchers, civil servants and government officials—and providing opportunities for sharing (both ways) with individuals and in groups of every size.

### Summary of a general assessment strategy

Figure 2.4 summarises the steps involved in an indicator project. The central role of stakeholder-participants is emphasised as is the iterative nature of many of the steps in the process. The project begins with a defining vision and ends with the sharing of significant findings. In the final events, but also through all the steps of the project, one can search out ideas for plans of action from individuals in the field and from policymakers, who will support those who are more directly involved in agriculture.

## 2.4 Quantification and Scaling of Indicators

By definition, indicators have a quantitative aspect, a value expressed as a number or a series of numbers, sometimes in graphical or other pictorial form. Once indicator values are obtained by some type of measurement process, the question arises as to whether the raw numerical value should be used in its original form or be converted into some other scale.

### Indicators using the original data

Using original data assumes that the raw numbers collected in the field or in a survey can be employed directly in the form in which they were obtained. In some cases, two or more sources of original data are combined in order to clarify the significance of the information.

- Some types of indicators are simple and are immediately understood by everyone without further manipulation or conversion of the numbers. For example, literacy would commonly be measured in percentage terms; it would be clear to anyone that 80 per cent literacy is better than 40 per cent literacy. Furthermore, in the case of literacy there is an obvious poorest value or lower limit (zero per cent), as well as an almost universally accepted optimum, or goal (100 per cent).
- There are other examples of indicators that are equally clear but which have only limited meaning when presented as a single number. At the global level, measurements of the average yearly temperature are made as a means of giving evidence (for or against) global warming. The Celsius temperature scale, which is familiar to most people, is used. There is no worst or best value, but results can be compared with a benchmark value, such as the temperature in a particular year or the average value over a number of years. The plot shown in Figure 2.5 uses the

**Figure 2.5**
Changes in average global temperature between 1860 and 2000

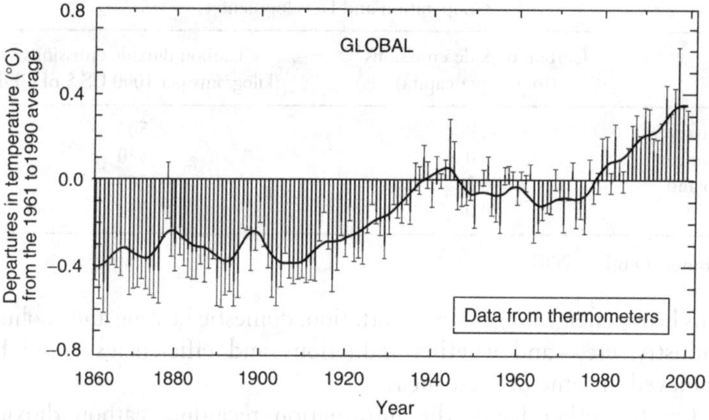

[Figure: Global temperature departures (°C) from the 1961 to 1990 average, with data from thermometers, showing values ranging from about −0.6 in the late 1800s to about +0.4 by 2000, y-axis from −0.8 to 0.8, x-axis from 1860 to 2000.]

Source: IPCC (2001), p. 107.

average temperature over the past 140 years as the benchmark. Clearly, in this instance, it is the trend of temperature variations and the magnitude of the difference from the benchmark value that are significant. A single temperature value, in the absence of any reference value or trend, carries very little significance.
- Other types of indicators present information that is less familiar, at least to the general public. Again, the indicator data become more meaningful when they are used to show differences at various locations or to show changes over time. Where quantities are presented, such as amounts of materials consumed or released, it is usually preferable to express the quantity in relative rather than absolute terms. If carbon dioxide emissions are being measured, these could be reported relative either to the population or to the gross domestic product (see Table 2.2).

These two ways of presenting emissions data are both useful, but their significance differs. The 'emissions per capita' information indicates that citizens of the USA are heavy users (relative to the other countries on this small list) of energy that is derived from the combustion of fossil fuels. This information could point to many other sustainability-related questions, such as ones regarding the purpose of

**Table 2.2**
Carbon dioxide release in selected countries (in 1998) as a pressure indicator of possible climate change effects. Data from the Organisation for Economic Cooperation and Development

|  | Carbon dioxide emissions (tonnes per capita) | Carbon dioxide emissions (kilograms per 1000 US $ of GDP) |
|---|---|---|
| Turkey | 2.9 | 500 |
| France | 7.0 | 340 |
| Poland | 8.2 | 1300 |
| USA | 20.0 | 740 |

Source: O'Brien (2001).

this heavy energy usage—transportation, domestic heating and cooling, industry, etc.—and whether reductions and efficiencies could be achieved in some of these areas.

On the other hand, the information regarding carbon dioxide release and GDP shows that Poland has an especially large value for its emissions per economic output, and this points to inefficiencies in the various sources of wealth generation that can release carbon dioxide (such as industry and agriculture) in that country.

## Elimination of the scale effect by taking ratios

The three examples just described each present data on different scales. The percentage and temperature scales are very familiar to most people. It is less common, however, that people would attach significance to information about amounts of carbon dioxide release unless it is used in a comparative fashion as described above. Elimination of units and scale is one means by which the comparative aspect of information can be made clearer. Elimination of scale is also necessary if one wishes to compare or combine various indicators calculated from a mix of different types of data sources. Because each source may provide information covering a unique range of values, and sometimes with different units, comparisons become difficult unless the separate indicators are transformed into some common form. How, for example, could you combine two types of climate information—one giving temperatures (degrees Celsius) and the other rainfall (mm of precipitation)—into indicators that have the same units. One way of eliminating scale is by taking ratios, and this can be done in several ways. To illustrate this, we can use as an example information about water availability for agriculture in a semi-arid situation that is supplemented by a limited amount of irrigation from a canal system.

In this case, over a 10-year period, the total amount (in mm) of water available from both rainfall and supplementary irrigation per year was 638, 720, 489, 595, 672, 640, 826, 720, 661 and 525 mm. The average value of water availability during this time was 649 mm and the variation, expressed as standard deviation,[1] was 98 mm. To eliminate the millimeter scale, several transformations can be used.

- Division by the average value leaves all the individual results in the same order, with the same relative standard deviation, but in a dimensionless form (i.e., a form with no units).

$$\text{New value} = x_i / x_{avg}$$

The set would then become:

0.983, 1.11, 0.753, 0.917, 1.04, 0.99, 1.27, 1.11, 1.02, 0.807

This method makes very clear whether results are higher or lower than the average, but it makes no statement about which values are considered to be desirable and which ones are unacceptable. Note also that there is no clear trend in values, with two or three low and high values appearing in both the first and last half of the data set.

- Division by the standard deviation converts results into a set with a standard deviation of unity.

$$\text{New value} = x_i / s$$

In this case the set would be:

6.51, 7.34, 4.98, 6.07, 6.86, 6.53, 8.42, 7.35, 6.74, 5.36

When using this method, the relative position of all results remains unchanged, but the bias of scale is eliminated.

- A method similar to this involves taking a ratio of the difference from the mean for each value against the standard deviation:

$$\text{New values} = (x_i - x_{avg})/s$$

The set becomes

−0.112, 0.724, −0.163, −0.551, 0.235, 0.092, 1.81, 0.724, 0.122, −1.26

Here, the scale bias is removed, but the relative position of each value has been altered.

## Indicators converted to a common scale (normalising data)

An alternative way of dealing with information that comes in a variety of scales and units is to convert the raw values into a new set of numbers with a common (usually dimensionless) scale. In some cases, qualitative data is assigned a numerical value within the scale as well. As with some transformations involving ratios, this can result in a set of simple and useful numbers that enable comparisons over time or with data obtained in other places. This process is sometimes referred to as normalising the data. In one sense, the choice of scale is not important, although scales such as 0 to 100 (per cent) or 0 to 10 are widely understood and acceptable in many situations. Because there may be considerable uncertainty in much of the data that one obtains in a sustainability study, it is probably wise not to establish too broad a range of possible values, as this could imply a level of accuracy that is often not justifiable. In this respect, a scale from 0 to 10 may be more acceptable than 0 to 100. In subsequent sections of the book, we will often make use of the 0 to 10 scale.

The method of using a common scale is particularly appropriate for situations where there are definable poorest and optimal values—numbers we shall call goalposts. These values set the upper and lower limits of the scale.

There can be several kinds of situations encountered in generating a set of scaled values.

## Assigning a numerical score to qualitative information

Qualitative attributes can be given a numerical score. For example, using a scale of 0 to 5 in a sustainability study, the following generic definitions could be applied to qualitative assessments of some activity or process:

- <1    unsustainable in all respects
- 1–2   approaching unsustainable conditions
- 2–3   partially sustainable
- 3–4   sustainable in most aspects
- 4–5   highly sustainable

When good quantitative data are available, on the other hand, using a limited scale range such as 0 to 5 may require rounding off numbers that are in some cases obtained from very reliable measurements, and could therefore obliterate subtle but significant differences between systems.

### Converting various data types to a common scale

In the examples developed below, we make an arbitrary choice to employ a scale of 0 to 10, and we report results to an accuracy of one decimal place. In reporting and using the calculated results, however, in some cases the decimals would be rounded off to the nearest digit.

Defining data values that are assigned a scaled value of 0 and 10 is an important and difficult first step in the scaling process. To illustrate some possible scaling manipulations, a good starting point is to work with the case of literacy, for which we have suggested the 'obvious' choices of 0 per cent as poorest and 100 per cent as best values. Using these boundaries (which we will call goalposts) to convert to the common scale, per cent values are divided by 10. A village whose inhabitants are 65 per cent literate would rank a scaled indicator value of 65/10 = 6.5.

Other indicators that are derived from data not based on percentage can likewise be converted to the same (or another) common scale. Average domestic water use per person in an African village might range up to 40 litres per day. This largest value could be chosen as the upper goalpost, with 0 as the lower one. Usage of 30 litres by an individual would then correspond to a scaled value of (30/40) × 10 = 7.5.

By converting to a common scale based on values with defined upper and lower limits, the process of comparing results from vastly different types of information is simplified. Without this, the user would be required to have clear background knowledge about the range of values that could be expected for each specific type of measurement.

### Selecting values for indicator 'goalposts'

In both the examples just described, the lower and the upper goalposts were, respectively, zero and a maximum value defined by the nature of the data. For the literacy case, we stated that it was obvious that the goalpost values should be 0 and 100 per cent respectively.

But is our choice of limits really obvious? The 0 and 100 per cent values could be the most reasonable choices in some situations, such as a country emerging from an historical situation, when there were no schools and the literacy rate was very low, into an age where education is available and more and more people are learning to read and write. In other situations, however, educational facilities have been widely available over many generations, and the literacy rate has been at least 80 per cent for the past 50 years. In this latter situation, reducing the number of illiterate people from 10 per cent to 5 per cent, probably a major accomplishment, would show up within the 0 to 10 indicator range as an apparently small change from 9.0 to 9.5. In this situation, it might be better to choose 80 per cent as the lower goalpost since it is the smallest percentage literacy in recent history. The range then runs from 80 to 100 per cent and individual values (given the symbol p) would be scaled as follows:

$$\frac{p - 80}{100 - 80} \times 10$$

Therefore, for the 90 per cent literacy situation, the indicator would be

$$\frac{90 - 80}{100 - 80} \times 10 = 5.0$$

Likewise, with the revised calculation, the 95 per cent figure has a scaled value of 7.5. By changing the lower limit and the range of values, we have caused the indicator to point to an apparently much greater improvement in performance in the area of basic education than was shown using the broader percentage scale.

It must be emphasised that the choices made here are arbitrary ones, but arbitrary does not mean that such choices should be made carelessly. As a public information tool, the 80 to 100 goalposts create a more sensitive indicator, one that responds more visibly to small changes (hopefully, improvements). It is important that the goals for defining indicators should be kept in mind in each specific situation. It is equally important that the methodology used in any study be made very clear so that policy makers and the public are fully able to understand and interpret the reported data.

There are many other examples of the types described above. Suppose we wish to scale the values of crop yields. In the first instance, definitions of good and poor values (and therefore the range) will vary from crop to crop, and will also depend on each local situation, including the environmental conditions that apply. For rice grown in a situation where irrigation is plentiful, excellent productivity might be defined as a yield of 9 tonnes per hectare. This yield would then be assigned an indicator score of 10. (Incidentally, any occasional higher yields would also be assigned the same maximum scaled score of 10, not a value above the maximum.) Again, the lower limit could be set as a complete crop failure—0 tonnes per hectare—or, alternatively, some minimal value, say 2 tonnes per hectare, might be arbitrarily selected. Other crops would be scaled in a similar manner but with their own unique set of limits. The scale for sugarcane is one that would be very different from one for rice or other grain crops. In the case of sugarcane, a realistic maximum value might be 100 tonnes per hectare.

Finally, rather than using goalpost values that are selected in advance of obtaining data, on the basis of previous knowledge and/or by arbitrary but carefully thought-out choice, ratings can be scored relative to the complete set of values obtained in the survey. This after-the-fact method involves setting out an ordered array of all the responses for each item, and then assigning a qualitative or quantitative value to individual measurements based on their position within the ranking. For example, on a scale of 10, results that are placed within the top 10 per cent of all values would achieve a score of 10, those within the eightieth to ninetieth percentile would be scored at 9, and so on.

## Taking weighted averages

As with other assessments of human activities, measuring agricultural sustainability can be done at various levels, from that of the individual farm up to measurements taken for entire regions (usually those sharing common features, such as the prairies of North America).

Even at the farm level, the data that are collected in the field and used in developing an indicator are frequently based on a series of measurements. This will inevitably be the case when a larger region is being studied. For example, crop yield information, such as that which has been described in the previous section, might require data about a particular crop in a given year, which includes information from many farms. Averaging such data should always be done by weighting the individual yields according to the area of land where that yield was obtained. To take a simple example, suppose that there are five farms growing sorghum in a given area with the following outputs.

| Farm size (ha) | Yield (kg/ha) |
|---|---|
| 25 | 2,400 |
| 30 | 3,200 |
| 20 | 2,200 |
| 10 | 4,100 |
| 32 | 2,300 |

A simple average of the yields gives a value of 2,840 kilograms per hectare (kg/ha). In this average, all the yield values have been considered as contributing equally to the average. The amount of land on which each crop was harvested differs significantly, however, and should be taken into account to give a meaningful average for the entire region.

To do this, the more correct method of averaging involves multiplying each yield value by the corresponding field size, summing these and dividing by the sum of all the areas. In this way, the average is adjusted to give greater weight to the larger fields.

Such a weighting procedure can be represented as:

$$\frac{\sum S_i Y_i}{\sum S_i}$$

In this case, the weighted average value is:

$$\frac{25 \times 2400 + 30 \times 3200 + 20 \times 2200 + 10 \times 4100 + 32 \times 2300}{25 + 30 + 20 + 10 + 32}$$
$$= \frac{314600}{117} = 2690 \text{ kg/ha}$$

The weighted average is somewhat lower than the simple average, reflecting the fact that the particularly high (4,100 kg/ha) yield, having been obtained on a small area of land, has therefore influenced the result to a smaller extent through the weighting process.

## Inverting data

For some measurements, small values are more favourable than large ones. When this is the case, it is sometimes possible to invert the question. Instead of asking, 'What is the percentage of land under monoculture?' it may be preferable to ask, 'What is the percentage of land under multiculture?' This way, a high percentage leads directly to a large indicator score.

Modifying the question is not always a simple matter, nor is it always desirable. In these situations a simple manipulation of the data can be carried out. A case in point is incidence of disease, e.g., disease associated with the use of chemical pesticides. For this issue, the optimum would be that there is no incidence of disease and this defines the lower goalpost as zero. The upper goalpost will be chosen from a defined 'most unfavourable' situation. For example, if over a period of years there have been between 300 and 600 admissions per year to hospitals in a given district that have been attributed to pesticide poisoning, a poorest value base of 600 could be set, with the range of possible values then falling between 0 and 600. In a given year when there were p cases of poisoning, to invert the data the scaled indicator value is calculated as:

$$\frac{600 - p}{600 - 0} \times 10$$

If there were 350 cases, the indicator would then take a value of

$$\frac{600 - 350}{600} \times 10 = 4.2$$

If the number of cases were reduced to 200 in subsequent years, the indicator would then rise to a value of 6.7.

Note the difference between the formula used here and the formula used for direct scaling in the literacy example.

## Measuring against 'benchmark' values

In the example just given, arbitrary decisions were made with regard to defining what can be called the 'poorest value' (in that particular example, the upper goalpost) in the calculation. Along with setting results for one or both the goalposts, it is sometimes necessary to compare the measured indicator values with one or more standard values that may differ from situation to situation. We will refer to these standard values as 'benchmarks'.

A good example of this comes up when comparing water use in different settings. It is always desirable to strive for efficient water use everywhere, but the degree of efficiency required for sustainability will depend on the available fresh water supply. For this reason, an availability benchmark should be established for every individual case, and intensity of water use should be expressed with relation to that benchmark value. In considering sustainability of water use, the standard usually chosen is the total available renewable freshwater resource, including inflows from neighbouring areas (or countries, depending on the level of the assessment). The indicator is then generally expressed in percentage terms.

Consider the following pair of countries:

| Per capita water use per year in the 1990s | |
|---|---|
| Canada | 1,600 cubic metres |
| Korea | 610 cubic metres |

| Water use as per cent of total renewable resource | |
|---|---|
| Canada | 1% |
| Korea | 35% |

The message from these two pairs of indicators is clear. From the first set, we see that Canadians are heavy users of fresh water, for agriculture, industry and domestic purposes. The second set, however, shows that because Canada is blessed with extensive fresh water resources in the form of lakes and rivers, and also because of its small population, the proportion of the renewable resource used each year is very small. The situation is reversed in Korea—a smaller degree of exploitation per capita, but a much larger stress on the total available supply. Clearly, the absolute amount of water used is not the important criterion to predict sustainability, but rather the amount relative to available resource.

Another example of an application of benchmark values is in situations where yields of different crops are to be compared. One crop may have an intrinsic ability to produce a higher yield than another and the optimum yield that could be expected of each would be designated as a benchmark value. Actual yields can then be expressed as a fraction or percentage of the benchmark. The fraction is then a measure of how close the land (and associated management practices) comes to achieving its productivity potential.

For example, consider a hypothetical situation involving two crops, A and B.

|        | Potential yield | Actual yield (tonnes per hectare) | Fraction |
|--------|-----------------|-----------------------------------|----------|
| Crop A | 8               | 6                                 | 0.75     |
| Crop B | 5.5             | 5                                 | 0.9      |

In this example, the actual yield of crop A is better than that of crop B, but expressed as a fraction, using the potential yield as a benchmark, crop B has proved to be more productive. In comparing yields of the two different crops, in some ways this comparative measure gives a better estimate of productivity.

## Combining individual indicators to produce a composite index

We have noted that information obtained within any study on sustainability, whether through measurements in the field or in the laboratory, is subject to manipulation or treatment—processing that is meant to simplify and clarify large quantities of data. One goal of the treatment process is to end up with a smaller, more manageable data set. Sometimes the goal is to produce a single number that attempts to incorporate all of the information into what is usually referred to as an index. Figure 2.6 illustrates the flow of information from the primary set to the possible ultimate index value.

The triangle in Figure 2.6 shows that a sustainability study begins with the collection of (often large) amounts of information. The primary information is processed by standard methods such as collating and averaging survey data, or evaluating analytical measurements by calibration with standard samples. At this second stage, the processed set essentially contains all the information required in the study. This complete set is itself useful for detailed investigation of specific features

## 96 Agricultural Sustainability

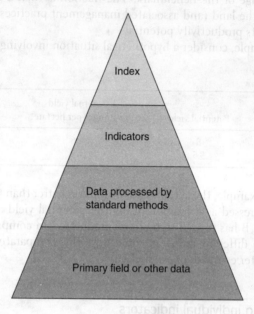

**Figure 2.6**
Information triangle showing the flow of information from primary data to final index

of sustainability and is essential as the basis for further manipulation into indicators. It should always be kept available for later checking of the validity of the indicator values and to answer questions that may be posed by persons who are using information contained in the indicators.

The next step is to convert the processed data into indicators by methods that have been discussed here, but even after this manipulation there could be a large number of indicators, often giving a mixed message about the degree of sustainability. It is sometimes useful to combine or aggregate the various indicator values into a single index or a very small number of integrative indices. We described some of the advantages and disadvantages of this aggregation process in an earlier section.

How is this combining done? The simplest way is to take an average of the individual indicator values, assuming that they have all been normalised to the same scale. For example, if a social index, expressed as a percentage, combining information regarding nutrition, health, education, and access to decision-making might be made up from indicators such as these:

## Sustainability Indicators    97

Percentage literacy                                            43
Percentage of children 12 and under attending school           87
Percentage of population receiving an adequate diet            62
Percentage making use of local clinic                          32
Percentage voting in the local elections                       79

A social index calculated as average value of the above data would be calculated as:

$$\text{Average \%} = \text{Social index 1} = 61$$

An alternative way of calculating the average, sometimes used in statistical studies, is by the Root Mean Square (RMS) method.

$$\text{RMS Average \%} = ((x_1^2 + x_2^2 + x_3^2 + x_4^2 + x_5^2)/5)^{1/2}$$

where the $X_i^2$ values are the squares of the individual indicator values. These are all summed, averaged, and the square root taken. In the present case:

$$\text{RMS Average \%} = \text{Social index 2} = 64$$

In both the calculations of averages, without being explicit about it we have assumed that each of the individual indicators should have equal weight in determining the overall index value. While the assumption has the advantage of not appearing to impose bias on the result, it may in fact do just that. In our example, the indicator about use of a local health clinic could be considered less important than the others, in the sense that a clinic may exist, and lack of use may not indicate lack of availability.

We might then choose to weight each indicator according to its perceived importance. For example, we could (again arbitrarily) assign a weight of one to four of the indicators and of 0.5 to the one related to health clinic usage. The calculation is now modified slightly using the weights, $W_i$:

$$\text{Weighted average \%} = (W_1x_1 + W_2x_2 + W_3x_3 + W_4x_4 + W_5x_5)/(W_1 + W_2 + W_3 + W_4 + W_5)$$

In the general case, the weighted average is $\Sigma W_i x_i / \Sigma x_i$

Weighted average % = 1×43 + 1×87 + 1×62 + 0.5×32 + 1×79)/4.5

$$\text{Social index 3} = 64$$

Likewise, the RMS average can be calculated using weights. The general formula for this is:

$$\text{RMS weighted average \%} = (\Sigma W_i x_i^2 / \Sigma x_i)^{1/2}$$

For our example,

$$\text{RMS weighted average \%} = ((1 \times 43^2 + 1 \times 87^2 + 1 \times 62^2 + 0.5 \times 32^2 + 1 \times 79^2)/4.5)^{1/2}$$

*Social index 4 = 67*

It can be seen that the weighting operation has discounted the value of one of the indicators, in this case the one with the lowest value. Therefore the new calculated averages are somewhat higher than in the unweighted cases. What is most essential is that decisions about weighting or not should be made very clear, in order that users of the aggregated indicators or index can make decisions with full knowledge of the basis of the index values with which they are working.

One point that needs to be made again and should be clear from this discussion of combining data is that there are both advantages and disadvantages associated with using data from various levels of the information triangle. The lower end of the triangle contains all the information, but this is often in a form that is not very accessible or easily interpreted. As we move toward the apex, the information becomes simplified and focused, but this is at the expense of detail that is sometimes essential, and it can be biased by decisions made by the creator of the indicators and indices.

There are other methods of treating and aggregating data that have been recommended for various situations, but all of them follow the general principles outlined here. The Human Development Index is an example of an index based on a compendium of indicators that has achieved global prominence.

## The human development index

One of the most thoroughly researched attempts to gauge human development, especially in its social and economic dimensions, is the index, known as the Human Development Index (HDI), developed by the United Nations Development Programme (UNDP). Based on a vast compendium of data collected around the world in many categories, three categories were identified as providing the essential information needed to define a simple yet comprehensive index of human development. These were indicators describing three issues: health, education and economic status. The indicators for each category were evaluated individually on a scale of 0 to 1, using the following general formula:

$$\text{Specific indicator} = \frac{\text{Actual value} - \text{Minimum value}}{\text{Maximum value} - \text{Minimum value}}$$

The maximum and minimum value are again arbitrarily chosen and set as goalposts for each indicator.

Once the three specific indicators were calculated, the HDI was determined by taking a straightforward average of the three.

HDI = 1/3 (health indicator) + 1/3 (education indicator) + 1/3 (economic status indicator).

## Health

A considerable collection of health-related data is available from many sources, including the World Health Organisation. Data for incidence of various diseases, for availability of health care facilities, for public expenditures on health and for mortality at or shortly after childbirth were considered for use in developing a health indicator. In the end, however, the most recent five-year average for life expectancy at birth was chosen as the single indicator sufficiently robust to provide an integrative assessment of human health.

For health, the life expectancy maximum and minimum goalpost values were chosen to be 85 and 25 years respectively. As an example, for Nigeria, where the average life expectancy at birth is 51.5 years, the health index is found to be 0.44.

$$\text{Health indicator} = \frac{51.5 - 25}{85 - 25} = 0.44$$

## Education

To measure access to education and success of the educational process, two indicators were chosen and then combined into one.

Adult literacy rates are widely available, and these were used as one part of the education indicator. The second part was derived from measures of the combined primary, secondary and tertiary gross enrolment in educational institutions. Gross enrolment ratios were calculated by dividing the number of children enrolled in each level of schooling by the number of children in the age group corresponding to that level. Although this measure has many deficiencies, such as differences in age ranges corresponding to different levels of education in various countries, it was considered to be the most satisfactory indicator from amongst the readily available data for assessing educational opportunities.

For both adult literacy rate and combined gross enrolment ratios, percentage values were used and goalposts set at 0 and 100 per cent. The two individual values were separately calculated and then combined to create an education index. Combining was done with a weighting of two-thirds to adult literacy and one-third to gross enrolment. Using the example of Nigeria, where adult literacy was 62.6 per cent and gross enrolment 45 per cent, the education index was calculated as follows:

$$\text{Adult literacy indicator} = \frac{62.6 - 0}{100 - 0} = 0.63$$

$$\text{Gross enrolment indicator} = \frac{45 - 0}{100 - 0} = 0.45$$

Education indicator = 2/3 (adult literacy indicator) + 1/3 (gross enrolment indicator)
$$= 2/3 \times 0.63 + 1/3 \times 0.45$$
$$= 0.57$$

Note here the arbitrary but clearly defined choices that were made in coming up with a good assessment of the educational status of a country.

### Economic status

Gross domestic product per capita expressed in US dollars was chosen as the measure of economic status. Because measurements in different parts of the world involve different currencies and different methods of classification and aggregation, results may not be strictly comparable. Nevertheless, the data are readily available and are considered to be a broadly acceptable measure of average economic status of individuals within a country.

To calculate the index of economic status, the GDP per capita is adjusted by taking a logarithm of the value. This has the effect of spreading out the indicator values where otherwise there would be many clumped together near the low end of the scale. The maximum and minimum goalposts are set at US $ 40,000 and US $ 100. Once again, we use Nigeria (GDP per capita value = $ 853) as an example:

$$\text{Economic status index} = \frac{\log(853) - \log(100)}{\log(40000) - \log(100)} = 0.36$$

Finally the overall Human Development Index (HDI) is taken as the average of the health, education and economic values.

$$\text{HDI(Nigeria)} = 1/3\ (0.44) + 1/3\ (0.57) + 1/3\ (0.36) = 0.46$$

Thus, the human development index for Nigeria is determined to be 0.46.

This widely publicised and important index provides a good case study illustrating how individual indicator values are calculated and scaled, and also shows a simple method of combining three indicators to come up with a single composite index. Note again that there are a number of places where arbitrary decisions are made: the choice of categories where measurements will be made, the data used for assessment within each category, the numerical method used to calculate the individual values (including choosing appropriate maximum and minimum values as goalposts), and the way in which the individual values are combined to provide a composite number.

The result is a single, simple value ranging from 0 to 1, a value that points to the status of human development in each country. As with all indicators, this one value serves simply as a 'grade on a national report card'. A particular value, say a low one, leads one to investigate further as to the reasons for that value. Is progress poor in all the categories, or is one particularly low individual result pulling down the average? And once general problem areas are identified, one is able to delve further into detailed causes.

## Other indicators used to measure human development

Many other measures of human development have been proposed. Most of these make use of more data than is required in the HDI. For example, Qizilbash (2001) proposes an 'Index of Poverty and Well-being' that can be applied to nations. It is made up of six components:

- People not expected to survive beyond age 40 (%)
- Adult illiteracy (%)
- Underweight children under age 5 (%)
- Proportion of population with access to sanitation (%)
- Combined school enrolment ratio (%)
- Consumption ($ US per capita)

For each of the individual indicators, countries are ranked, with the country having the poorest score (note that a poor score may be high or low, depending on the indicator) being assigned a rank of 1. The next poorest score is given a rank value of 2 and so on. For all the countries, the sums of rank values are then put in order using the same system, and an overall rank value is then calculated. Clearly this scoring method is useful for comparative purposes, but it is difficult to interpret the significance of any individual result in an absolute sense.

An important point to note is that the Qizilbash system proposed for measuring human development, while different from the HDI in the details of indicator choice and method of aggregation, is similar in its selection of issues to be addressed. Both the HDI and the Index of Poverty and Well-being describe human development by including as key features indicators for health, opportunities for education and financial resources. The fact that both approaches categorise human development in similar ways indicates that they share a common conceptual overview.

Likewise, as we consider ways to measure agricultural sustainability, a primary goal should be to develop an overview of the issue that can gain broad acceptance.

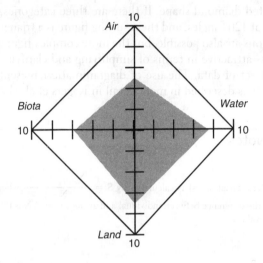

**Figure 2.7**
'Diamond' used to illustrate four components of environmental sustainability. Shaded area is a diagrammatic representation of the extent of sustainability for a particular example.

## Diagrammatic ways of combining information

Where several indicators have been calculated, each showing one aspect of sustainability that is under study, it is possible to combine and report values using a diagram, so that the individual components are all visible at one time.

Consider a hypothetical case where indicators of environmental quality have been determined under categories of air, water, land and biota. Using a common scale, wherein values of 10 are considered optimal within each category, the values for a particular situation can be plotted in a way that illustrates how close the situation is to the ideal. Suppose in this hypothetical environmental study, scaled values for the categories are: air, 4.5; water, 7.2; land, 6.8; and biota, 6.0. The plotted values, including the ideal, form 'diamonds' as illustrated in Figure 2.7.

The perimeter lines represent the ideal, while the shaded area within these corresponds to the actual situation for a particular case. The extent of shading provides an illustrative (but not quantitative) view of the degree to which the situation approaches sustainability.

The degree to which the shaded area is skewed in one direction or another depicts the strength or weakness of each category.

The example given here is for a four-component indicator set, meaning that there are four axes, and the polygon to be plotted takes on a distorted diamond shape. If there are three categories, the axes are plotted at 120° angles, and the resulting figure is a triangle. Higher order polygons are also possible, but the more complex figures become less and less attractive in terms of simplifying and clarifying a multi-component set of data. The use of diagrammatic representations of sustainability is described in more detail in Rogers et al. (1997).

## Notes

1 The standard deviation (s) is calculated by $S = \sqrt{\dfrac{\sum(x_i - x_{mean})^2}{N-1}}$ where $x_i - x_{mean}$ values are the difference between individual and average values. N is the number of individual values.

# 3 | Indicators for Assessing Agricultural Sustainability

Using our framework definition and description of agricultural sustainability, we have set out six categories (see Chapter 2.2) to be considered in order to make a comprehensive assessment of any agroecological system. We have labeled these as productivity, stability, efficiency, durability, compatibility and equity. In various interrelated ways, the six categories encompass features of the environmental, economic and social tripod aspects of sustainability. Using the categories as headings, we can create and examine possible indicators that are applicable directly or in modified forms in different agricultural situations.

Many of the indicators suggested in the following sections are ones that the authors developed for the study of agricultural sustainability in four agroecosystems in the Tungabhadra Project area in Karnataka State in South India (vanLoon et al. 2001a). Other persons have developed different but comparable sets of indicators for different situations. In Chapter 4, some of these studies are reviewed. As will become clear, there is no definitive set of categories that can be applied universally to all situations. Each agroecosystem, with its own geographic, social and economic uniqueness, requires a corresponding unique approach to assessment. Yet, many of the broad principles outlined here can be widely applied, and as more and more assessment work is done it may be possible to develop a broadly acceptable set of basic indicators.

Some of the proposed indicators are described in detail, while in other cases, where development of the specifics depends on knowledge of each particular situation, only the background principles are provided. Frequently, it will be necessary to make judgment calls in choosing the most appropriate indicators and then in setting goalpost values or in defining a scale. In the specific examples provided here, we have adopted a 0 to 10 scale for most of the indicators. This is, of course, arbitrary, and others who are evaluating sustainability may prefer to operate within a different range of values.

## 3.1 Productivity

*Productivity*   *For the needs of the farm family as well as to satisfy global food requirements, any sustainable agricultural system must be capable of producing good yields.*

Productivity is, of course, the immediate and primary goal of any agricultural enterprise. While it may appear obvious, let us restate and make clear that there are two essential reasons why it is important to ensure that abundant crops are produced on land around the world. One is the necessity to grow sufficient food for the individual farm family, for the community and for the world at large. There is only a limited supply of high quality agricultural land in the world (see Figure 2.2). In the past several decades, more and more of the Earth's terrestrial environment has been taken from its natural state and converted to use for agricultural purposes. Increasingly, the new farmland is marginal and has limited agricultural potential. Because of this, large amounts of high-energy material and labour inputs may be necessary in order to modify the natural ecosystem into an imposed agricultural one (see Figure 1.6). At this same period in history, while marginal land is being taken over for agriculture, formerly good agricultural land is being lost at an accelerating rate due to erosion, desertification, salinisation and other soil degradation processes. It has been estimated that close to 15 per cent of the world's land mass has undergone some degree of degradation in the 50 years up to 1990, and about 40 per cent of the degraded land (6 per cent of the total) is considered to be highly degraded. The extent of degradation continues to increase. Gains in food production must therefore come largely from increases in productivity on already existing farms. And gains are essential if we are to feed the world's growing population.

The other reason why high productivity is an essential feature of sustainable agriculture is to provide a 'good living' for those involved in agriculture. Although there may be debate about what is necessary for life and what constitutes 'good living', we would insist that, besides providing sufficient food for an adequate diet, agriculture as an occupation should also make available some measure of comfort, access to education and health care, as well as opportunities for relaxation, recreation and cultural participation.

No one denies that agricultural productivity is an essential and laudatory goal. Some have argued, however, that while productivity is a most important issue, it should be considered as an issue that is independent of the subject of sustainability. In other words, agriculture should be assessed in two parallel but separate ways. We believe that this separation is problematic. By separating the two issues there is a temptation to think in terms of trade-offs—and very often productivity, as a response to obvious and urgent needs, would win out at the expense of sustainability.

We argue, therefore, that it is a mistake to remove productivity from the sustainability equation. Rather, productivity must always be taken as *an intrinsic component of sustainability*. Any twenty-first century agricultural system that claims to be sustainable but produces crops at only a marginal level is in fact not sustainable. Such systems can support neither the global needs of food and fibre, nor the individual needs of farmers as a source of food and income. On the other hand, any production system that results in excellent yields and profits, but gradually or even precipitously destroys the resource on which it is based, is equally unacceptable and unsustainable. It is for these reasons that we include productivity as an intrinsic category in a comprehensive assessment of sustainability.

In measuring productivity, the fundamental issues centre around how much crop can be produced on a given area of land. For most individual crops (we are not referring here to a multiple-cropping system) there is a primary component and a secondary one. For cereals, for example, the grain itself is the primary component while the straw is secondary. The yield of the two products together can be referred to as biomass yield: the total quantity of plant material that is produced during the growing season. While the secondary component is usually considered to be the less important of the two products, it should never be thought of as a waste material. There are several possible uses to which it may be put, the main ones being as a soil amendment, as cattle fodder, or as a fuel. Later, in looking at issues of efficiency and compatibility, we will discuss the relative merits of these three uses and other potential uses of straw and other secondary materials.

The most fundamental measures of productivity are based on the total amount of biomass produced. This *gross productivity* is defined as the mass (or volume) of new plant material produced by photosynthesis within a given area over a specified time. Typical measurements are made in units such as kilograms (kg) or tonnes (t) per hectare (ha) per year (y), although locally common units such as bushels per acre are also in frequent use. Gross productivity is difficult to measure. While the plants making up the crop are undergoing photosynthesis to synthesise organic matter, they also respire, releasing some of the synthesised material back into the atmosphere as carbon dioxide and water. Still more of the plant material is lost through decay, consumption by herbivores and other causes. The difference between photosynthetic production and respiratory degradation and other losses is termed the *net productivity* or *yield*, and this is what is determined when one measures the total mass of crop produced in a given area over a set period of time. For ease of measurement, it is common to include only the above-ground parts of the crop (except in the case of tuber crops) in determining net productivity or yield.

## Productivity potential

The potential maximum net productivity of a section of land depends on several factors, including the crop being grown, nature of the soil, temperature, duration and intensity of sunlight, and amount and timing of water availability. There is therefore no single goal that can be set for all situations in terms of highest productivity. Natural ecosystems themselves show a broad range of net productivity values, from more than 30 tonnes per hectare (t/ha) per year in tropical rain forests through around 5 to 10 tonnes in temperate grasslands and savannas to much less than 1 tonne in arid lands. For different crops in various parts of the world, there is likewise varying potential for maximum yields. This is evident in the long-term average global yields for the world's six most important crops (see Table 3.1). These are average values, and the range can be rather large. Note also that in many settings, double or even triple cropping is possible and this augments by factors of up to 2 or 3 the total productivity of the land per year.

Note that the ratio of primary to secondary component of the crop is variable, depending on the plant, but in many cases the ratio is quite close to 1:1. In the absence of specific information for a particular

**Table 3.1**
Global long-term average net productivity of the six most important crops. *Primary* refers to the grain or tuber that is the human food product, while *secondary* refers to the remaining above ground parts of the plant

| Crop | Primary t/ha | Secondary t/ha |
|---|---|---|
| Rice | 2.5 | 2.5 |
| Wheat | 1.9 | 2.9 |
| Maize (Corn) | 3.0 | 4.5 |
| Sorghum | 1.0 | 1.2 |
| Cassava | 9.0 | 6.3 |
| Sugarcane | 46.0 | 9.0 |

**Source**: Data from Smil (1993), p. 203.

situation, the simple ratio of unity is a reasonable approximation that can be used when necessary. While net productivity—the sum of values in the two columns of Table 3.1—is the true measure of the ability of the agroecosystem to produce biomass, most often yields are calculated on the basis of the primary product alone.

In any case, because of the intrinsic variability in potential productivity, it is usually not appropriate to directly compare yields of different crops unless they are referenced to benchmark values (as described in Chapter 2, using a hypothetical example of two crops). To do this, it is necessary to make a choice concerning the most suitable benchmark.

This can be illustrated using actual yield data in the following way. The global, Chinese and Indian average yields in 2001 for a number of key crops are given in Table 3.2.

Clearly, even though both sugarcane and rice are frequently grown in a similar well-irrigated region, one would not wish to compare the yields of these two crops directly. The conclusion of such a simple comparison would be that, in all the instances shown in the table, sugarcane is more than 10 times as productive as rice. This type of comparison of yield among two or more crops essentially provides confirmation of the already available information about the photosynthetic efficiency and biomass productivity of different parts of different plants.

By using benchmark values for comparison, one can measure performance against a standard or ideal. Possible benchmark values could be the best value ever obtained, the average global yield (over a

Table 3.2
Global, Chinese and Indian average yields for a number of important food crops in 2001.

| Crop | Average yield in tonnes per hectare | | |
|---|---|---|---|
| | World | China | India |
| *Cereals* | | | |
| Barley | 2.60 | 2.32 | 2.03 |
| Maize | 4.43 | 4.93 | — |
| Millet | 0.78 | 1.76 | 0.74 |
| Oats | 2.12 | 1.62 | — |
| Rice | 3.91 | 6.34 | 2.96 |
| Sorghum | 1.36 | 3.13 | 0.74 |
| Wheat | 2.73 | 3.83 | 2.74 |
| *Root crops* | | | |
| Cassava | 10.5 | 16.0 | 25.9 |
| Yams | 9.53 | — | — |
| *Pulses* | | | |
| Lentils | 0.82 | 1.41 | 0.75 |
| Soybeans | 2.33 | 1.77 | 0.93 |
| *Other crops* | | | |
| Plantains | 6.23 | — | — |
| Sugarcane | 65.2 | 77.0 | 70.6 |

Source: Data taken from the Food and Agriculture Organisation (FAOSTAT data, 2003, at http://faostat.fao.org).

selected period of time), average yield within the country of interest, or some value based on an augmented version of an average.

For example, 1.5 times (or some similar figure) the global average yield could be taken as representing an attainable yield that is highly productive. Comparisons can then be made with this 'ideal', using ratios expressed directly or on a scale of 0 to 10 or as percentages. Using this system, if one were comparing productivity of wheat in India and China, the global average wheat yield of 2.73 t/ha would be converted into a target yield of $2.73 \times 1.5 = 4.10$ t/ha. The ratios for the two countries then give indicator values as follows:

$3.83/4.10 = 0.93$ (9.3 on a scale of 0 to 10), for China,
and $2.74/4.10 = 0.67$ (6.7 on a scale of 0 to 10), for India.

This approach, where calculations of indicators are always made in a common, systematic manner, also allows for comparisons between the productivity of different crops. In India, lentils and soybeans can be grown in similar areas at the same time of year. A comparison of productivity of these two crops would be done as follows:

For lentils, global average = 0.82, with target value of
$$0.82 \times 1.5 = 1.23 \text{ t/ha}$$
Indicator value, expressed on a 0 to 10 scale = $(0.75/1.23) \times 10 = 6.1$
For soybeans, global average = 2.33, with target value of
$$2.33 \times 1.5 = 3.50 \text{ t/ha}$$
Indicator value, expressed on a 0 to 10 scale = $(0.93/3.50) \times 10 = 4.0$

In this calculation, it is clear that while the yield of soybeans was substantially higher than that of lentils in 2001, its calculated productivity index is smaller. This suggests that production of the lower yielding crop, lentils, actually came closer to the benchmark value than did the production of the intrinsically higher yielding soybeans.

In cases where a specific yield value is higher than the benchmark 'ideal', the indicator would normally be assigned the maximum value in the range. In the examples above, these maximum values would be 1 as a ratio, 10 in the scale 0 to 10, or 100 on a percentage scale.

Some interesting modifications of this approach for calculating productivity are described in Chapter 4, where a detailed case study in Central America is described.

In each of the above calculations, arbitrary choices regarding benchmark values have been made. If a rationale for the choices and the detailed methodology are clearly described in reporting results, it becomes possible for persons doing other studies in different settings to convert yield information from these sites into equivalent indicator values that allow for comparisons to be made.

### Productivity indicators

With this background, indicators can be designed that measure the productivity of one or more crops grown on a defined area of land using particular agricultural management practices. Where we deal with all the crops in a particular area, we refer to the combination of land, crops and management practices as an agroecosystem.

In the first instance, productivity will be gauged by measuring total biomass and primary product. When measuring productivity, the yield from a single field in a single year, while important to the farmer, has very limited meaning in terms of overall sustainability. Variations in climatic and other natural and management conditions frequently result in yield variations on the same land of 20 per cent or more on a yearly basis. Some sort of average is then called for. For example, an integrated value, based on farms within a common agroecosystem and/or covering a number of years, could be used to serve as a reliable indicator of productivity. Several state indicators are then recommended for assessment in different situations. To illustrate the calculations, consider a small region where three crops are grown with the following amounts harvested:

|       | Area (hectares) | Primary harvest (tonnes) | Secondary harvest (tonnes) | Primary yield (tonnes/hectare) |
|-------|-----------------|--------------------------|----------------------------|-------------------------------|
| Maize | 35              | 182                      | 270                        | 5.2                           |
| Wheat | 27              | 103                      | 131                        | 3.8                           |
| Oats  | 13              | 40                       | 62                         | 3.1                           |

### Crop productivity indicators

**P1a/b (Primary Product Yield or Conventional Yield):** For a single crop, the conventional yield is calculated simply as the mass of primary product per unit area. The standard units are tonnes per hectare. All over the world, yearly values of crop yields have been determined and tabulated for individual situations and as averages. This indicator is useful for comparison of trends in a particular area, for comparisons between areas, and for comparisons of new varieties of crops or different management systems. In addition to calculations for a single field, average yields within a region and/or over a number of years can be determined. Where this is done, such as when making comparisons for a particular crop grown in different agroecoregions, weighted mass values are calculated according to the area of each field.

The Primary Product Yield indicator can be calculated for one field or for a single farmer (P1a), or for many farmers, crops and/or years using a weighted average (P1b).

$$P1a = Y_p = M_p/a \quad \text{(for a single crop or farmer)}$$

$$P1b = \frac{\sum Y_{pi} \times a_i}{a_t} \quad \text{(for an agroecosystem)}$$

# Indicators for Assessing Agricultural Sustainability 113

where $M_p$ is the total mass of crop, $Y_p$ is the yield and $Y_{pi}$ is the yield of primary product grown within area $a_i$ and $a_t$ is the total area.

For the region described in the example, the primary product yield is:

$$P1b = (5.2 \times 35 + 3.8 \times 27 + 3.1 \times 13)/75 = 4.3 \text{ t/ha}$$

The indicators P1a and P1b can be compared to a benchmark and scaled to give a more generally-applicable measure of primary product yield (P1bs).[1]

$$P1bs = \frac{\sum \frac{Y_{pi}}{Y_{pbi}} \times a_i}{a_t} \times 10$$

$Y_{pbi}$ represents the benchmark yield: values for individual crops, such as 1.5 × global average for each.

In the example, using benchmark yields given in Table 3.2:

$$P1bs = (((5.2/6.65) \times 35 + (3.8/4.10) \times 27 + (3.1/3.18 \times 13))/75) \times 10 = 8.7$$

**P2a/b (Biomass Yield):** The Biomass Yield is similar to the Primary Product Yield, but is a measure of the average total biomass production per area of land of a principal crop (P1a) or of all crops (P1b) within a defined agroecosystem.

$$P2a = M_m/a = Y_m \quad \text{(for a single crop)}$$

$$P2b = \frac{\sum Y_{mi} \times a_i}{a_t} \quad \text{(for all crops in the agroecosystem)}$$

where $M_m$ is the total (primary + secondary) biomass of the crop, grown on area 'a', and $Y_m$ is the total biomass yield. $Y_{mi}$ is a yield value for an individual farmer growing a crop on area $a_i$. Thus $\Sigma(Y_{mi} \times a_i)$ equals the total mass of all crops grown within the agroecosystem of total area $a_t$.

Depending on the purpose of the indicator, yields could be measured for one year or averaged over a period of several years. In the latter case, a good assessment of the general productivity of the agroecosytem, independent of the unique weather conditions of any one year, is obtained.

Like the Primary Product Yield, in order to scale the values (0 to 10), these two indicators can also be expressed as a ratio against a benchmark biomass yield value ($Y_{mb}$) using, for example, some multiple of the global average biomass productivity for the specific crop or for that type of agroecosystem. For example, the scaled value (P1as) of P1a is:

$$P2as = \frac{\sum \frac{Y_{mi}}{Y_{mb}} \times a_i}{a_t} \times 10$$

The biomass index includes the yield of both primary and secondary products. Although this indicator is not commonly used by farmers for typical grain crops or other food crops, it is often used as an indicator of productivity in agroforestry systems. It can, however, be applied to any crop, in which case it allows for comparison of net photosynthetic conversion ability.

Where two or more crops are grown on the same piece of land during a one-year period, the total productivity of the land can be calculated using P1 or P2 type calculations, but including yields from the two (or more) cropping seasons during the year.

Yield measurements on primary and secondary products are however not the only possible measures of productivity. Nutrient yield is also an important consideration. The principal nutrient of interest is usually protein. Most of the nitrogen content of plants is in the form of protein, so analysis of the plant material for nitrogen becomes the means by which the amount of protein is estimated. A factor of 6.25 is used to convert from nitrogen to protein. This means that grain containing 2.1 per cent nitrogen can be estimated to have about 13 per cent protein content.[2] All cereals contain substantial quantities of protein, and cereals are a major source of this nutrient in many human and animal diets. Other crops, particularly pulses, contain even larger proportions of plant protein. Because of its nutritional importance, productivity of protein could then be another calculation that one would wish to make.

**P3 (Nutrient Yield):** The Nutrient Yield indicator is the weighted average yield of a particular nutrient in the primary product of the principal crop on one field or within a defined agroecosystem, expressed as a ratio against a benchmark value, such as the global average for that crop. Most often the nutrient of interest is protein, and protein yields can either be expressed as mass of protein per unit area or as mass of nitrogen.

Table 3.3
Qualitative indicators for various crop properties

| | | | | |
|---|---|---|---|---|
| Protein Content of the grain produced (%) | | | | |
| poor <12 | normal 12–16 | | excellent | >16 |
| Total Protein Yield (kg/ha) | | | | |
| poor <400 | normal 400–600 | | excellent | >600 |

For other crops grown for specific purposes, similar qualitative assessments can be made.

| | | | | |
|---|---|---|---|---|
| Fibre Production (kg/ha) | | | | |
| poor <800 | normal 800–1500 | | excellent | >1500 |
| Gross Energy (GJ/ha) | | | | |
| poor <5 | normal 5 to 8 | | excellent | >8 |

$$P3 = Y_n/Y_{nb}$$

where $Y_n$ is the yield of a particular nutrient, and $Y_{nb}$ is a benchmark value yield for that nutrient.

In some cases, rather than reporting quantitative indicator values, textual values that give a relative qualitative measure of the desirable qualities of the product might be reported. One recommended method for qualitative scaling is given in Table 3.3.

## Multicropping

Each of the indicators discussed till now relates to measures of productivity in fields where a single crop is grown, although calculations of combined values for a number of single-crop fields have also been described. Single cropping is, of course, a common practice in many agricultural systems. Monoculture on a field scale has the advantage of relative ease in carrying out the various agronomic operations: ploughing, cultivating, sowing, weed control and harvesting. In most cases, highly mechanised systems require that a single crop be planted in a given area. Monoculture agriculture can also be productive in the conventional sense of providing optimal yields of typical field crops. A broad field of wheat or rice, well maintained, weed-free and showing lush growth, gives an impression of excellent agricultural practice. Evaluation of gross productivity in these situations is a relatively straightforward process that involves measuring the weight of the crop, and calculating yield based on the known field area.

With all its advantages, growing single crops over large areas can be a contributor to serious environmental problems. Having a very low level of crop diversity is especially problematic in terms of insect and disease control. The single species and variety of plant may become an attractive host for specific pathogens, while the lack of other species means that there is no appropriate niche area where neutral or predator species can reside in close proximity. As a consequence, in monoculture agriculture there is a temptation to rely extensively on chemical control methods to cope with the insect or disease problem. Another issue is that the sole crop will have defined nutrient and other requirements for growth. In order to maximise yields, it is often considered necessary to supply these requirements largely or exclusively by readily calibrated chemical means.

In contrast to the monoculture system just described, other agricultural systems are quite different. Mixed farming in West Africa, for example, often involves small plots where five or six crops are planted together, in individual rows or clumps or sometimes intermixed, apparently at random. Cassava, yams, sweet potatoes, and a variety of vegetable crops are frequently grown together in small fields surrounded by natural vegetation. In the northern part of Karnataka State in South India, a type of planting called *Akri* is common in rainfed areas of the arid Deccan Shield. In the *Akri* method, various legumes such as pigeon pea, *mung* and cowpea are planted along with castor bean in a single row. This row is separated from others of the same mixed type by four rows of sorghum. There is a large measure of security in this process. Sorghum is the principal grain grown in the area and, besides providing the farmer with a good supply of the staple food crop, the mixed row yields a small harvest of important legumes.

Because of the diversity of crops it is usually more difficult to determine productivity in a multicropped system. One method is to calculate the Land Equivalent Ratio.

**P4a (Land Equivalent Ratio):** This is an intercropping indicator for two or more crops grown together in a given area:

$$P4a = \Sigma Y(P)_{Ci}/\Sigma Y(M)_{Ci}$$

In this calculation, the denominator term ($\Sigma Y(M)_{Ci}$) is the sum of the yields for each of the crops when they are grown in monocultures. The

**Table 3.4**
Example of yields obtained for two crops grown separately or together. In the multicropping system, the total area is 1 hectare and the ratio of land area for *mung* beans to sorghum is 1:4

|  | Mass obtained in multiculture for a one-hectare plot (in kg) | Yield in multiculture (kg per hectare) | Yield in monoculture (kg per hectare) |
|---|---|---|---|
| *Mung* beans | 320 | 1600 | 1500 |
| Sorghum | 560 | 2800 | 2500 |

numerator term $(\Sigma Y(P)_{Ci})$ is the sum of corresponding yields when grown together in the multiculture system. A value of 1 indicates that the productivity of the multicropped system is the same as that for the crops grown individually. A value greater than 1 shows greater productivity and less than 1 shows reduced productivity. Clearly, in order to do this calculation it is necessary to have comparable information about yields for the two crops grown both alone and in an intercropped system, under otherwise identical conditions.

Consider a case where sorghum and *mung* beans are grown together in a one-hectare plot, with one row of beans interspersed between four equally spaced rows of sorghum. The land devoted to these two crops is then equivalent to 0.2 and 0.8 hectares respectively. In Table 3.4, the first and second columns respectively give the mass and yields obtained in the intercropped system, while the third column gives equivalent yields for these crops grown alone in the same area.

For this intercropped system, the value of the land equivalent ratio is then equal to:

$$(1600 + 2800)/(1500 + 2500) = 1.1$$

This value indicates that in this example there is a measurable positive advantage in terms of yield in the intercropped situation. According to the calculation, 10 per cent additional land would be required to obtain the same yield when the two crops are grown as two separate monocultures. The increase during intercropping seen here could be due to the positive effects of the legume on nitrogen content of the soil, and of the sorghum in providing shade, which in some instances would enhance germination of the low-growing intercrop. Besides yield, there are other advantages to multiculture agronomy, advantages that will be discussed more fully in the section on compatibility. Even where the Land Equivalent Ratio is less than 1, these advantages may be sufficient to favour the multiculture option.

## Cash crops

With respect to issues of sustainability and its measurement, cash crops can be of special interest. Cash crops is a term that has been used in a wide variety of ways. In the context of sustainability, we describe such crops as having some or all of the following properties:

- In most cases, they do not benefit the cultivator or the region in terms of basic food security (i.e., provision of essential food energy and protein) but may provide a lucrative income to the farmer, and are sometimes a source of valuable foreign exchange.
- They often command a higher price for their products than traditional crops, making them attractive in terms of monetary returns.
- Many are crops that have been introduced recently into a particular area.
- They are grown in specific areas, either where the climatic conditions are suitable for maximum yield of the crop or where market forces have been strong enough to influence the traditional cultivation practices of local farmers.
- In many situations, cash crops are subject to specialised and more intense high-input agricultural practices, partly due to the motivation of farmers to maximise the cash output of the crop. This motivation stems from the fact that, unlike traditional crops, cash crops are usually not food crops, or at least not essential food crops. This fact adds pressure on the cultivator to maximise profits in order to be able to afford food items that were traditionally grown on the farm but now must be purchased.
- Cash crops are frequently highly risk-prone in terms of variations of yield and market price.

In light of these traits, cash crops often influence and exert pressure on the surrounding environment, the local economy and the prevailing social conditions in a more profound and vivid way as compared to the more traditional food crops.

Consistent with this definition and description are non-food crops, such as fibre crops including cotton, hemp and jute, and products grown as industrial resources, like some forms of cassava. Flowers are another cash crop having high value and increasing economic importance in some regions. Also included in the cash crop listing are many horticultural crops—fruits and vegetables, cane or beet sugar, spices, tea and coffee. Clearly, while these can contribute in important ways to the local food basket, they are frequently grown as items for sale in other national regions or for export.

In some cases, cash crop productivity can be measured in traditional ways; for example sugarcane yield is usually determined in tonnes per hectare. In other cases, however, diverse and innovative measures are required, each depending on the characteristics of the specific crop.

### P5a (Yield of Cash Crops):

$$P5a = \text{number of units/area}$$

Where the crop production cannot reasonably be measured in mass units, then the number of items will be used. An example is the production of flowers, where the unit may be individual flowers, or bunches of a particular size or containing a set number of stems.

In assessing productivity of cash crops, it will frequently be most appropriate to measure productivity in monetary terms, since yield measurements such as those described above do not allow for comparisons with traditional food crops. Monetary measures are appropriate for traditional crops as well.

### Monetary indicators of productivity

Referring again to the second objective in achieving good productivity—the goal of supplying a good income to the farmer—there is a need to establish indicators that will measure the monetary factor. One strategy for designing such indicators is to measure the *income productivity* related to a particular crop. This would provide information about the economic potential of that crop in the given region. Alternatively, the indicator could measure the profitability of all the crops on a whole farm or group of farms. In calculating income productivity, net values (crop value less the cost of production) should be used. Once again, individual farmers may wish to obtain this information on a yearly basis for their own individual holdings, but a study of sustainability within defined areas would require that the information be averaged from many farms, and possibly over a number of years.

For a given region, the indicators will measure the average income productivity. A separate but also important issue is the distribution of incomes. Distribution falls under the equity heading, and will be dealt with later.

The following indicators are recommended as measures of income productivity:

**P6a/b (Income Yield):** Income Yield is a recommended measure of income productivity. It is defined as the weighted average net income per hectare for crops within a defined agroecosystem.

$$P6a = I_n/a \qquad \text{for a single crop}$$

$$P6b = \Sigma(I_{ni} \times a_i)/a_t \qquad \text{for a defined agroecosytem}$$

where $I_n$ is the net income and a is the area. $I_{ni}$ is the net income per hectare for each crop grown on a field of area $a_i$.

Where more than one crop can be grown each year, the calculation can be done either over one season or over one year. In the latter case, the calculation requires determining the total income from the two or more crops grown on a given piece of land during a year.

Scaling of monetary productivity indicators requires making some arbitrary choices, based on experience in that region, of upper and lower limits, and then developing a scale to cover the range of values. In most cases, one would use local currency units.

$$P6as = (I_{ni} - P)/(B - P) \times 10$$

where B and P are the best and poorest monetary productivity for the agroecosystem.

An example or calculation of Income Yield as described by P6as for a sorghum crop grown in India might be as follows:

Yield = 1.7 t/ha, selling price Rs 2.5/kg, gross value = Rs 4,250, costs = Rs 750, net value = Rs 3,500

Optimum yield (arbitrary) = 3.5 t/ha, gross value = Rs 8,750, B = optimum net value = Rs 8,000

Poorest value (arbitrary) = 0.7 t/ha, gross value = Rs 1,750, P = poorest net value = Rs 1,000

The scaled value is then:

$$P6as = (3,500 - 1,000)/(8,000 - 1,000) \times 10 = 3.6$$

For local and national internal use, the crop value will be expressed in local currency units, while for international comparisons conversion to a standard monetary unit such as US dollars can be done. It is common that only the primary product will be employed in the calculation, but where the secondary product has commercial value, this can be included as well.

Monetary indicators depend only partially on biophysical productivity and are variable, depending on changing market conditions. An advantage of monetary indicators over those based on amount of material produced, however, is that they allow for comparison of different (even totally different) crops (e.g., banana and sugarcane, which can be grown under similar agroecological conditions). Therefore such indicators can provide economic justification for decisions regarding choice of optimum crop in a given agroecosystem.

**P7a/b (Total Net Income):** Total Net Income is defined as the weighted average total net income based on crop production for farmers within a defined agroecosystem. In some cases, it is useful to report this as a ratio to the per capita national income (or some other value) as the benchmark value.

$$P7a = \Sigma I_{ni} \quad \text{for a single farmer}$$
$$P7b = \Sigma\Sigma I_{nij}/j \quad \text{for farmers within an agroecosytem}$$

where $\Sigma\Sigma I_{nij}$ = the total net income from all crops for all farmers. There are j farmers within the agroecosystem.

Scaling of total net income for a farmer is carried out in the same manner as described above for a single crop. In the Total Net Income case, the poorest value might be chosen to be the regionally defined poverty level but the optimum value would be a more arbitrary choice. Clearly, in this calculation the size of the land holding is a major factor that determines the income productivity. As a consequence, this indicator incorporates issues of productivity and equity.

The Total Net Income for a single farmer in scaled form is then:

$$P7as = (\Sigma I_{ni} - P)/(B - P) \times 10$$

B and P are the best and poorest total net income values chosen for this system.

As seen in these indicators, productivity calculated by any of the methods suggested above can be determined for a single crop on one field, or it can be averaged for one or several crops for a number of fields and/or farmers, and it can be averaged over time. Of course, any space-based averaging should be weighted by area. When income productivity is calculated for many crops within a given area, benchmark productivity values are required to normalise results for the individual crops.

When income productivity measures are calculated for different agroecosystems, they provide a reasonable measure of economic equity between regions.

---

**Box 3.1**
**Strategy for assessing productivity**

- The scope of the assessment should be determined—number of crops, area and time to be covered.
- Choose at least one indicator from those suggested for measuring *crop productivity* (P1 to P5) and one for *income productivity* (P6 or P7). Note that indicators identified with the letter *a* are generally appropriate for assessment of productivity of an individual crop while indicators identified with the letter *b* are to be used for several crops or within a defined ecosystem. Where several indicators within either the crop productivity or income productivity sub-categories are measured, an average value of these indicators is calculated.
- Scale each of the chosen indicators to a common scale. A scaling process, based on a 0 to 10 scale, has been suggested for some of the basic indicators.
- Calculate the average of the two sub-category values. Use an unweighted average, which then incorporates an assumption

*(Continued)*

*(Continued)*

> that crop yield and monetary value are equally important. If there is a good reason for weighting individual values, assign reasonable weights, reporting reasons for assigned values.
> - Where appropriate, calculate other productivity indicators in order to add additional supporting information to the study. Note that all the indicators suggested here are state indicators. Maintaining a time series of productivity data is a particularly useful exercise.

The productivity indicators described here are recommended as components of a package of indicators that can be employed in assessing, in a holistic manner, the sustainability of agriculture in a particular region. There are other related but somewhat different uses to which the same indicators can be put.

One of these additional functions is for comparing productivity between different agroecosystems with a view to identifying factors that lead to good productivity. There will be expected environmental factors including an assured water supply and appropriate temperature and solar radiation regimes. One expects (and usually finds) that productivity is very largely determined by environmental conditions—climate, soil etc.—in the setting where the farming is done. But equally important will be the connections between management practices and good productivity. Are good yields dependent on substantial inputs of non-renewable resources, such as chemical fertilisers and synthetic pesticides? How does productivity compare between areas where conventional tillage is used and those where no-till technology is practised?

There have been many case studies designed to investigate productivity using different farming systems; in particular, comparisons have been made between crops grown using conventional agriculture, which makes use of chemical fertilisers and pesticides where necessary, and production following lower-input, more organic practices. One cannot claim that in all instances the latter types of agricultural management systems are as productive. Especially in the short term, it is frequently possible to achieve better yields using high-chemical input methods, but this is not always the case. There are a good number of studies in diverse situations around the world which provide solid evidence that comparable and sometimes even better productivity is achievable with low-input management systems.

In experiments carried out in Tigray in 1997–98 (Edwards 2002), various crops were grown in parallel in unimproved control plots along with crops in heavily composted plots and crops in chemically fertilised plots. The composted plots gave greater yields than the control plots in every case, by factors of 3 to 5. Moreover, the yields on composted plots were more often than not better than on the chemically fertilised plots. The range of yields in composted plots as compared to chemically fertilised were (for the common grains): wheat, between +20 per cent and 0.2 per cent; barley, between +9 per cent and −0.5 per cent; maize, between +7 per cent and −21 per cent; and finger millet, +3 per cent.

Altieri (2001) gives examples from South and Central America of improved yields associated with the adoption or re-adoption of organic methods. In Mexico, yields of coffee increased (often doubled or even more), after rejecting chemical fertilisers and instead using composting along with contour planting and terracing. In Guatemala and Honduras, major increases in maize yields were achieved after ploughing in green manure and using other soil conservation methods. Yield increases of up to 2.5 times were achieved in a wheat/maize rotation by Brazilian farmers, again by using green manures and other cover crops.

The kinds of issues raised by these findings lead us to the category of stability.

## 3.2　Stability

> *Stability*　It is necessary that the high level of productivity be maintained over an indefinite period of time. This requires that the quality of the resources on which production is based also be maintained and even enhanced.

Stability in crop production describes the ability to maintain a good level of productivity over an extended period of time. In terms of time, we must keep a long-term view in mind, ranging beyond decades to centuries—in this respect we take seriously the intergenerational aspects of sustainable development. With this definition, the only true measure of agricultural stability is to observe production systematically over many years in a location where a single set of agricultural practices has been followed during that time. Controlled experiments of this type have been carried out in various countries. Perhaps the best known of these are the Rothamsted field trials being conducted for over a century at an agricultural research station in Harpenden, Hertfordshire, in southern England (Leigh and Johnston 1994).

While such experiments have been underway in various parts of the world and provide unique and highly useful information, the length of time required is a serious drawback, and it is only possible to investigate a limited number of agroecosystems in this manner. As an environmental issue, the situation regarding measurement of stability is in some ways analogous to that surrounding the subject of global climate. In both cases, the definitive experiment is to follow the yield productivity or average world temperature over many years, while carefully assessing the practices that may affect these parameters. In both cases, after these experiments have continued for decades it is possible that a conclusion would emerge indicating that serious problems had already arisen—in the case of global climate, that current practices had compromised the Earth's climate so that it was a much less habitable place for humans with consequent untold suffering. Likewise, with agriculture, widespread famines would ensue if a long-term experiment indicated that certain common practices had led to lowered productivity. Clearly, it is impractical to think of basing decisions solely on the basis of long-term agronomic experiments.

Leaving aside the Rothamsted-type field trials, at some other locations (individual farms, or data collected within a limited area) there is information about the yield of one or more crops on the same land over a span of perhaps 10 or 20 years. This information too is useful, but one or two decades may be too short a time to reveal evidence of slowly declining productivity. Furthermore, care must be taken in evaluating evidence of apparently stable or increasing production. What are the measures being used to maintain good yields—are these measures themselves sustainable? It is these kinds of issues that need to be incorporated in a comprehensive determination of sustainability.

A related way of measuring stability (and one with similar limitations) is to keep a record of crop yield variability. This is done by tabulating crop yields and then calculating the average over a given period of time along with a statistical factor to measure the variability in yields. The standard deviation is one such factor. The significance of this kind

## Table 3.5
### Rice yields (tonnes per hectare) in a single field between 1976 and 1999 (hypothetical data)

| Year | Yield | | Year | Yield | Year | Yield | |
|---|---|---|---|---|---|---|---|
| 1976 | 7.3 | | 1983 | 6.8 | 1992 | 6.2 | |
| 1977 | 7.0 | | 1984 | 6.9 | 1993 | 7.0 | |
| 1978 | 6.7 | | 1985 | 7.1 | 1994 | 6.0 | |
| 1979 | 7.4 | SET 1 | 1986 | 7.6 | 1995 | 7.8 | SET 2 |
| 1980 | 7.4 | | 1987 | 7.4 | 1996 | 7.8 | |
| 1981 | 7.0 | | 1988 | 6.7 | 1997 | 6.8 | |
| 1982 | 7.5 | | 1989 | 6.8 | 1998 | 7.9 | |
| | | | 1990 | 7.5 | | | |
| | | | 1991 | 7.6 | | | |

of assessment can be made clear by considering the following data set for rice yields in an irrigated field over a 24-year period (Table 3.5).

The average yield in the first seven years (Set 1) is 7.2 tonnes per hectare and the variation expressed as standard deviation is 0.29 tonnes per hectare. Over the final seven years (Set 2) the corresponding average and standard deviation are 7.0 and 0.73 tonnes per hectare respectively. What these data suggest when comparing the final with the initial yields is that a small yield decline has occurred. This decrease is, however, far too small to be suggestive of any real change in productivity. On the other hand, the greater variability in yield, as indicated by a standard deviation that is almost twice as large in Set 2 as compared to former years, does give evidence that yields are more erratic and that additional inputs are perhaps required in order for them to be maintained at a good level.[3]

Having said this, it must be emphasised that great caution should be used in interpreting any data (including use of standard deviations of yield) that cover only a limited period of time.

For most situations, it is better and often necessary to measure stability by employing indirect methods. Indirect measurement factors consider changes in the nature and quality of the supporting resources, the natural capital needed for plant growth (soil, water, etc.) and other inputs. For example, in areas where soil erosion occurs to a significant extent, yields over a 10-year period may be quite stable, perhaps maintained by using chemical inputs to make up for loss of quality topsoil. However, measuring the rate of erosion (note that there can be both gains as well as losses of topsoil by erosion) over a relatively short time may be a good indicator of possible instability in the long term.

The two most fundamental resources for crop production are soil and water, and the maintenance of their quantity and quality that can be used as indirect measures of stability of crop production.

## Soil and water degradation

Because maintaining good crop productivity is a central feature of agricultural sustainability, it is essential that the basic soil and water resources remain available in ample quantity and good quality. In spite of this importance, vast areas of productive agricultural land are currently being lost to other uses or degraded through improper use and poor management. Some of the losses are associated with capture of land for use by industry and growth of urban areas. More than 75 per cent of persons in high income countries and 35 per cent in low income countries now live in cities and these percentages continue to increase substantially every decade. Estimates of amount of land devoted to industry, living space and other non-agricultural uses range from 22 hectares per person in India to 60 hectares per person in the United States. Unfortunately, it is all too common that the urban areas are built on some of the highest quality land since rich agricultural areas, often adjacent to rivers or lakes, have historically been the settings for establishment of major cities. To some degree, acting as a counterbalance against these losses, there is now increasing interest in promotion of urban agriculture—small and often richly productive gardens using private or public space for growing high-value crops such as vegetables and fruits.

Other losses of productive land are specifically associated with land degradation, both due to physical causes—wind and water erosion, waterlogging—and chemical causes—nutrient depletion, soil salinisation, acidification, and toxicity. In certain environmental situations around the world, such as in West Africa, one or more of these causes has led to the general phenomenon known as desertification.

The other fundamental resource required for productive agriculture is water, the availability and quality of which must also be sustained. Precipitation is the most basic source of water, but irrigation using water diverted from surface sources as well as extracted from groundwater reserves is also important—globally, upwards of 19 per cent of productive land is being supplied with these additional water supplies. As we noted earlier, approximately two-thirds of all the water that is extracted from surface and sub-surface sources is used for agriculture, an amount far greater than the requirements for industry and for domestic use.

Stability indicators are required in order to measure the continuing availability of good quality soil and water for food production.

## Soil quantity

In any situation involving the land, whether or not it is influenced to a significant degree by human activities, there are ongoing processes of both soil formation and soil loss. Formation is due to a variety of physical and chemical weathering processes, whereby the mineral material in the local rocks is converted into altered constituents that are finely divided and have a high and active specific surface area. Many of the physical and chemical processes of soil formation are greatly accelerated by microorganisms and the actions of plants and trees. New productive soil is also created in some areas through agents such as rivers that deposit silt in a delta.

Leaving aside situations like these, where large quantities of soil are brought in by major transfer activities, soil formation *in situ* is a very slow process. Even in the tropics, where chemical reactions proceed relatively rapidly, due to high temperatures and sometimes high rainfall, estimates of the rate of new soil formation are generally no greater than 1 tonnes per hectare (t/ha) per year. This corresponds to a depth of soil of something around 0.1 mm per year.[4] At this rate, it would take some 2000 years to develop a topsoil of 20 cm depth, a minimal depth needed to support productive plant growth.

Operating against formation processes are processes of loss, with transfer by wind and water being the most important factors. Sloping lands are especially susceptible to water-borne erosion. Losses can be highly variable, but under a well-managed soil they are very small. Moreover, the small amounts of soil carried away by wind or water may very likely be deposited in adjacent fields, with a net change of soil quantity at any point being apparently close to zero. However, there are many situations where exposed land is subject to major losses, much greater than can ever be made up by natural processes in a short period of time. Losses of 100 tonnes or more per hectare per year have frequently been recorded.

**Figure 3.1**
Observable features that indicate soil erosion: (a) Cross-section of a rill. Typical dimensions might be 10 cm width and 5 cm depth. (b) Pedestals are present around obstructions in the soil and indicate that soil has been lost over a broad area, leaving a stabilised portion as a remnant. (c) An armour layer is a layer of coarse aggregates overlying the intact soil, where finer material has been removed

In measuring soil quantity, we are therefore concerned with changes in the amount of nutrient-rich topsoil that can continue to support productive agriculture. In most instances, it is very difficult to measure changes in the depth of topsoil over a short period of time, and some indirect measures of losses are once again called for. Soil is removed from site by various types of erosion; evidence that erosion is occurring becomes an indicator of instability.[5] The evidence that we are looking for includes obvious accumulations of soil adjacent to trees, clumps of plants, or on gentle slopes. Dust in the air and muddy water in streams and rivers are further clear evidence of erosion. Erosion can be directed through large or small channels in the soil and these can be readily observed. It can also occur in a more diffuse manner. Sheet erosion, meaning that a consistent layer is removed over a large area, is more difficult to observe than the more localised types of erosion.

Within eroded fields then, we must look for indirect evidence of erosion. Rills, gullies, pedestals and armour layers on the soil are the kinds of visual observations that are most frequently made. These terms are defined as follows:

- Rills—small, narrow, irregular channels, generally having a roughly triangular cross-section (Figure 3.1a), are associated with erosion due to runoff from rainfall. They commonly occur on sloping lands where there is little or no vegetation.
- Gullies—like rills, but much larger and very obvious; often an extension of a natural depression in the land. Slumping of the soil banks adjacent to the gully widens and deepens it.

## Agricultural Sustainability

**Table 3.6**
Visual indicators of soil erosion

| Indicator | Wind erosion | Water erosion |
|---|---|---|
| Dust storms and clouds | × | |
| Sandy layer on soil surface | × | |
| Muddy (turbid) water after rainstorms | | × |
| Sediment accumulation in reservoirs | | × |
| Rills | | × |
| Gullies | | × |
| Deposits on flat lands or gentle slopes | | × |
| Accumulations of soil around obstructions | × | × |
| Pedestals | × | × |
| Armour layers | × | × |
| Exposed roots above soil surface | × | × |

Source: Adapted from Stocking and Murnaghan, (2001).

- Pedestals—the remnant of the eroded surface; a small tower-like structure usually formed around supporting material such as a clump of strongly-rooted plants or a stone (Figure 3.1b). Pedestals are one means by which evidence of sheet erosion can be observed.
- Armour layers—observed by the presence of coarse aggregates on the surface, overlying the finer soil material below (Figure 3.1b). They are evidence of removal of surface fines, again often involving sheet erosion.

Table 3.6 lists visually observable indicators of soil loss by wind and water erosion and indicates the associated type of erosion.

Some observable features can be roughly quantified to give an estimate of the soil loss in a given area. For example, consider an area of land criss-crossed with rills. Take a typical rill having dimensions 10 cm (0.10 m) wide and 5 cm (0.05 m) deep, and also having the usual approximately triangle-shaped cross-section (Figure 3.1a). This small channel is 3 m long and serves as a drain for a catchment area estimated to be 5 m long and 2 m wide. The volume of soil lost, based on the volume of the rill, is calculated to be:

$$\underbrace{0.5 \times (0.10 \text{ m} \times 0.05 \text{ m})}_{\text{cross-section area}} \times \underbrace{3 \text{ m}}_{\text{length}} = 0.0075 \text{ m}^3$$

The area drained is 5 m × 2 m = 10 m². Converting this single rill, which is chosen to represent a larger area, the loss of soil per hectare (10,000 m²) would be 10,000/10 × 0.0075 = 7.5 m³. Since each cubic metre of soil weighs approximately 1.3 tonnes, the mass of soil carried away can be estimated to be 7.5 × 1.3 = 10 tonnes (approximately). This calculation is based on a realistic situation and indicates the huge losses of topsoil (10 times greater than the estimated most rapid yearly rate of soil formation) that can occur due to erosion. The calculation is however a very crude estimate. You can see that defining the dimensions of a 'typical rill' serving a 'typical catchment' area is fraught with potential errors at every step in the calculation. Furthermore, we again ask the question, 'Where does the eroded soil go?'

**Ss1 (Erosivity):** This is an indicator measured by observation:

$Ss1s$ = qualitative measure of erosion having occurred

Unless one is willing to make a very detailed study, with careful measurements and estimates, for purposes of estimating soil loss in a general study of agricultural sustainability, it is probably better to use a purely qualitative rating scale, such as that suggested in Table 3.7. In keeping with our usual scaling practice, this table is based on a 0 to 10 scale, but limits description to four degrees of erosivity. In the proposed indicator ($Ss1s$), 10 represents minimal soil loss and is therefore indicative of a stable system, while 0 represents severe soil loss and a high level of instability.

The scoring system suggested for indicator $Ss1s$ can be modified as one becomes more familiar with the observations of erosion effects. Intermediate scores (e.g., between 3 and 7), can be assigned to situations where the levels of erosion appear to be intermediate to those described above. Assessments can be carried out on one field, or a composite value for the agroecosystem can be estimated by obtaining average observations within a larger area. Erosivity measurements can then be combined with other measures of stability to give an overall picture describing the possibility that the resources needed to sustain agriculture remain intact and of high quality.

## Soil quality

Over time there have been many reasons for studying soils, but two stand out as being of direct and practical importance to humans and

**Table 3.7**
Indicator Ss1s, describing scores associated with different levels of erosivity that can be observed in the field

| Score | Observation |
|---|---|
| 10 | No evidence of erosion of any kind: no structures in the field that show movement of soil, no 'dust' in the air on windy days, no turbidity in adjacent streams or drainage channels |
| 7 | Minor evidence of erosion as shown by one or more of the following: a small number of shallow rills, some evidence of downslope movement of soil, a limited number of superficial roots exposed, slight dustiness on windy days, minimal amounts of sediment in waterways after rainfall (especially heavy rainfall) |
| 3 | Clear evidence of erosion in one or more of the following ways: more extensive and prominent observations than noted in the previous category; some pedestals are evident, but of limited height. More frequent and deeper exposure of roots of plants and trees. |
| 0 | Severe erosion: extensive movement of soil as shown by deep rills (or even gullies) covering a significant portion of the soil surface. Accumulations of soil in the form of pedestals, 5 cm or more in height, and accumulations against stationary objects, considerable exposure of plant roots, subsoil exposed or close to the soil surface |

other living species. One of these is that soils are the principal plant growth medium and form the basis of agriculture and forestry. Especially over the last century, the subject of soil science has developed around this. Soil chemists have been particularly concerned with nutrient cycles and the relations between elements in the soil and their uptake by plants. They are also interested in other agronomic factors such as the interconnections between composition of soil materials, particle size, soil texture and the resultant soil physical properties. Soil science of this type—related to plant production—is a highly developed science with a body of knowledge that has increased in volume and sophistication over more than a century.

Another reason for studying soils is more recent and relates to the fact that soils play a major role as an environmental agent. Key links in the global carbon, nitrogen, phosphorus and sulphur cycles (as well as in many other cycles) involve soil chemical processes. Organic matter decomposition, nitrification, denitrification, phosphorous fixation and sulphide oxidation are just a few of these processes. There are two broad environmental implications related to such reactions. While soil chemical processes affect the nature and amount of elements that are released into water and the atmosphere, soils are in turn the locus of inputs from these other compartments of the environment, with the result that chemical changes occur in the material coming into the soil and in the soil itself. For example, rainfall chemical composition is altered when rain percolates through the soil, perhaps draining into rivers and lakes or maybe reaching the water table and becoming part of the groundwater reservoir. Soil properties are also altered by their encounter with the rain through these interactions.

Many soil reactions involved in the global element cycles have been occurring over long periods of geological time, although human activities in recent years have perturbed some of them to a significant extent. There are other specific types of chemical reactions that have only recently come to be played out in the soil environment. A good example is related to the application of pesticides in agriculture. Pesticides are used to control insects, weeds and pathogenic microorganisms on growing crops. These chemicals degrade over time and their movement and rate of degradation are in part determined through their interactions with the soil. Another example is the disposal of waste materials—municipal garbage, mine tailings, sewage sludge, sometimes even known toxic materials—in the soil environment. In other words, soils are an important environmental agent and a study of the environmental properties of soils is important along with studies of their agronomic properties.

In assessing agricultural stability, these two aspects of soil science come together. This is especially true in the context of maintenance of soil quality.

We see that soil is a critical component that plays diverse roles in regulating biogeochemical functions on the Earth. In the present context, in order to ensure continuing stable production of crops, it is essential that the quality of the soil be maintained or enhanced. It is for this reason that appropriate measurements of soil quality ought to be made and used as indicators of agricultural stability. But what do we mean by soil quality and what are the appropriate measurements? In order to develop suitable indicators, we must take into account the different functions of this resource and consider these functions with respect to the physical, chemical and biological nature of soil. The complex nature of the problem indicates that a single, simple indicator will not suffice for this purpose.

There are at least three issues that apply in considering the general characteristics of soil health (Doran et al. 1994):

1. The capability of the soil in supporting and maintaining a high level of plant and/or animal productivity over an indefinite period of time.
2. The ability of the soil to attenuate and at the same time prevent accumulation of environmental contaminants, so that they are not removed offsite into other areas or media, via water, soil movement or harvested crops.
3. The property that the soil does not adversely affect the health of plants or animals (including humans) in any direct or indirect manner.

In most agricultural situations, there are a large number of more or less standard soil measurements and tests that are routinely done (or at least can be done). What is required is to select from these a suite of tests which can taken together give a comprehensive picture of soil health. Ideally these tests should speak to the physical, chemical and biological properties of the soil, and they should be simple enough to be widely available in test facilities around the world. There would ideally be global consensus that enables comparison of results from standard tests, but it has frequently been observed that particular analyses are appropriate only within a limited range of environmental conditions. Earlier it was noted that there are a variety of nitrogen analyses that have been recommended for different applications. Another good example of the need for various tests is in the measurement and interpretation of data related to exchange capacity in soils. The standard tests that were originally developed for use with soils of the temperate regions of the world—soils that usually have a fixed negative charge on the clay minerals and organic matter that makes up the exchange complex—are not appropriate for the tropics and must be modified when applied to the variable charge soils commonly found there.

One set of basic and widely applicable indicators of soil quality recommended by Harris and Bezdicek (1994) for application to sustainability studies subdivides soil properties into physical, chemical and biological traits (Table 3.8):

The physical properties essentially relate to soil/water/air relations. A good soil is one that has the ability to retain water adequately so

**Table 3.8**
Soil physical, chemical and biological property quality parameters

| Physical properties | Chemical properties | Biological properties |
|---|---|---|
| Soil texture | pH followed as a time series | Microbial biomass in terms of carbon and nitrogen |
| Depth of soil and rooting | Electrical conductivity | Potentially mineralisable nitrogen |
| Soil bulk density and infiltration | Total organic carbon and nitrogen | |
| | Available nitrogen (ammonia and nitrate), phosphorus and potassium | Ratio of biomass carbon/total organic carbon |
| Water holding capacity | | Level of metabolic activity measured as soil respiration |
| Water retention characteristics | % of land covered by plant species with N-fixing associations | |
| | | Ratio of respiration/biomass |
| Infiltration | | Population density of key species in numbers per cm$^3$ |
| Water content | | |
| Soil temperature | | Earthworm activity |

that it does not dry out quickly. At the same time, it is sufficiently porous so that it drains well, minimising the possibility of waterlogging and making it readily workable. This balance is achieved when a soil has an intermediate texture, dominated neither by coarse material (sand) nor fine material (clay). Aggregation of soil particles assisted by organic material in the soil aids in maintaining good structure with adequate porosity.

The chemical requirements of a good soil are many and complex. The soil should be neither too acidic nor too alkaline, but should have a pH around a neutral value. The soil should not be host to a large accumulation of soluble salts. Most importantly, it should have a good supply of nutrients in a readily available form. The major nutrients, and the ones that should be measured in the first instance, are nitrogen, phosphorus and potassium. Minor and trace nutrients are also essential, but they may not always be measured in a rapid or basic survey of sustainability. While good soil properties can be achieved in a variety of ways, a particular feature that is favourable in many respects is the presence of organic matter in the soil. Organic matter is, to a limited degree, a source of nutrients. More important, it provides exchange sites that retain nutrients in a readily available form. Organic matter is also of benefit in enhancing good water retention capabilities, while adding to the porosity of the soil. Indicators that focus on each of these properties are required.

The soil biological quality parameters are field-based measurements whose values will depend on landscape, location, environmental and management conditions, and time of year. A soil with a dynamic community of organisms capable of decomposition and pest control is of higher quality than a soil with reduced numbers of fauna capable of only limited functions. In addition, soil organisms are important in nutrient cycling and in the maintenance of soil conditions such as structure and porosity. A decline in organism numbers is indicative of changes within the system and of potential changes in production. Thus soil organisms are useful indicators of the sustainability of resource use because they are directly affected by subtle changes in soil quality.

There are many options possible in deciding which soil microbial population measurements would be most sensitive as indicators of changing soil quality. One classification of sensitivity (Visser and Parkinson 1992) takes the following form:

- High sensitivity: populations of nitrifiers, *Rhizobium*, and actinomycetes; organic matter degradation rate, nitrification rate
- Medium sensitivity: populations of algae, bacteria, and fungi; soil respiration measured as carbon dioxide release or oxygen uptake; denitrification and ammonification rates
- Low sensitivity: populations of *Azotobacter*, total microorganisms, ammonifiers; aerobic nitrogen fixing capacity

Clearly, in measuring soil quality, it would be most desirable to select measurements of the most critical parameters and most sensitive processes if the choice is available. Note that many of these properties are measured by standard tests used in assessing the ability of the soil to support crop production, but they go beyond tests that measure soil quality *at the time of measurement* and are designed to indicate the possibility of problems that may occur *in the longer term*.

There are other recommendations for a minimum set of measurement parameters required to assess soil quality. Most of these sets have similar features, but each reveals biases and preferences of individuals related to their experiences in particular agricultural situations.

Table 3.9
Farmer recommended top ten measures of soil health given in rank order

| Rank | Property | Healthy soil | Unhealthy soil |
| --- | --- | --- | --- |
| 1. | organic matter | OM as high as possible, maintained by compost, manure etc. | lack of organic matter |
| 2. | crop appearance | green, dark, lush, dense tall, uniform | yellow, light green, stunted |
| 3. | erosion | soil stays in place | dust clouds, evidence of water erosion |
| 4. | earthworms | plentiful, visible after ploughing or after rain, holes and casings | not visible |
| 5. | drainage | drains properly, no ponding, dries out | saturated, ponding, drains too fast |
| 6. | tillage ease | one pass and ready, breaks up, smooth, mellow | needs more disking, pulls hard |
| 7. | soil structure | crumbly, loose, holds together | hard, lumpy, compacted falls apart, too light |
| 8. | pH | balanced, around 6.5 or 7 | too high, too low |
| 9. | soil test | up to recommendations, high, stays up each year | below recommendations |
| 10. | yield | better than average, consistent | lower than average, reduced |

Source: Paraphrased from Romig et al. (1995), based on interviews with Wisconsin (US) farmers.

## Farmers' opinions

A study (Romig et al. 1995) of how farmers assess soil health and quality involved consulting over one hundred Wisconsin (US) agricultural producers, to determine their perspective regarding what observable features are most important as determinants of soil quality. Many of their observations are applicable in diverse situations and provide a sound basis for prioritising parameters for stability assessments.

The selected top 10 soil health properties and some descriptive features associated with healthy and unhealthy soils are listed in order in Table 3.9.

In other parts of the world, documented or undocumented information from farmers can likewise provide invaluable information about essential features for high productivity in the local situation.

Assessment of soil health is not a simple matter, involving as it does issues that relate to soil physical, chemical and microbiological properties. Some of these are quantitative, but others are qualitative and must therefore be converted into simple scaled values when used as indicators. A further complication is that the magnitude and significance of these indicators can vary seasonally and depend on the fundamental characteristics of the soil resource as well as on the farm management system. It is therefore not possible to give universally acceptable optimum and poorest values that can be used everywhere for scaling results.

Accepting these limitations, we propose a set of state indicators here along with some general comments on how they can be evaluated. The set is not meant to be exhaustive, but applies to measurements of the three types of soil properties: physical, chemical and biological. Some of the indicators are ones that can be determined by farmers or by others through qualitative observations made in the field. Others require laboratory measurements to be carried out. Most importantly, we indicate the purpose of each indicator and its strengths and limitations; this will provide a basis on which to make choices, either of the indicators presented here, or of alternate measurements that are more suitable in other situations.

Because of the wide variability in individual situations around the world, specific details for quantification are in most cases not provided here. It is, however, recommended that the scale of measurement not be too finely divided, even when it can be based on accurately measurable analytical properties. If a 0 to 10 scale is used, it is best that it be subdivided into a small number of points, e.g., 0, 3, 7, and 10, as in the case of erosivity. 0 represents soil with very poor properties and 10 is soil with properties that are exceptionally good. As with many other aspects of sustainability, a time series of measurements provides much more information than can be gleaned from a single value. Increasing bulk density, decreasing content of organic carbon and nutrients are examples of changing properties that provide evidence of a decline in soil quality.

## Soil physical properties

Indicators to measure soil quality in terms of physical properties are based on qualitative observations in the field. Details regarding measurement and scaling depend on each specific situation.

**Ss2 (Soil Structure):** Soil structure refers to the manner in which individual particles are held together to form larger aggregates. Where few of these aggregates are present, the soil is said to have poor structure. A poorly structured soil, especially one with a substantial clay content, is heavy and resists downward movement of water. In some cases, a surface crust forms, inhibiting the flow of water into the soil column.

The formation of aggregates depends in part on the intrinsic textural properties of the soil but is also strongly influenced by the presence of organic matter, which has the ability to bind particles together into larger units. The spaces between the units allow for good drainage of water, minimising opportunities for waterlogging (with its attendant problems). At the same time, finer pores within the soil serve to retain water, thus preventing rapid drying to the wilting point. Stable aggregates also withstand the impact of raindrops, surface flow of water, and the erosive properties of wind. Aggregation is therefore a good measure of two things: the extent of water relationships with the soil, and the soil's ability to resist erosion. There are various measures that can be applied in assessing structure and its influence on soil properties. These include measures of water holding capacity, infiltration rate, dispersible clay, and observations of the surface and sub-surface appearance of the soil.

**Ss3 (Soil Bulk Density):** Soil bulk density, measured in grams per cubic centimeters ($g/cm^3$, a unit that is numerically equivalent to tonnes per cubic metre, $t/m^3$), can be determined in a variety of ways. A simple field measurement involves careful core sampling, removing and weighing the soil, then lining the hole with thin plastic and measuring the volume of water needed to fill the hole. The ratio of weight of soil in grams to volume of soil in cubic centimeters is then the bulk density. Bulk density values of about 1.2 to 1.35 $g/cm^3$ are characteristic of well-aerated surface soils while sub-surface soils typically have substantially larger values.

Bulk density depends on organic matter content, as most organic matter is light and porous, creating air spaces and thus lowering the soil density. Density also depends on degree of compaction, with compacted soils having bulk density values that are sometimes greater than 2 g/cm$^3$. Surface soils become compacted due to overuse of heavy equipment and this leads to poor aeration, poor drainage, increased erosion and poor root growth. No-till practices are primarily a response to problems arising from soil compaction.

A trend showing increasing bulk density over time is therefore an indication of declining soil quality, associated either with compaction or loss of organic matter.

### Soil chemical properties

There are a number of fundamental soil chemical properties that are commonly determined in soil testing laboratories. Where information from these tests is available, it can be used to establish indicators of soil quality in chemical terms.

**Ss4 (Soil Organic Carbon or Organic Matter):** Soil organic carbon or matter (in percentage terms) is determined using standard laboratory procedures involving wet or dry oxidation of the total soil. Because carbon content of soil organic matter is typically approximately 60 per cent, a factor of 1.7 is used to convert from organic carbon to organic matter content. For example, if a laboratory reports an organic carbon content of 0.77 per cent, the organic matter content is estimated to be 0.77% × 1.7 = 1.3%. For most essentially mineral soils, larger values are considered optimal, in the sense that organic matter promotes favourable soil structure that leads to good water retention/infiltration properties. Organic matter also provides exchange sites that favour nutrient retention and availability for plant growth. There is, however, no universal optimal value; tropical soils in arid regions tend to have much lower concentrations of organic matter than their counterparts in temperate areas. In a time series, declining organic matter content is generally a sign of reduction in quality. Organic soils such as those developed on peat deposits are a separate case.

**Ss5 (Soil Reaction):** The soil reaction, also called the pH value (measured in the lab using a soil slurry), is an intrinsic property dependent on the nature and amounts of parent organic and inorganic materials.

Nonetheless, pH is also affected by environmental and management influences. Acid rain and improper (usually excessive) use of nitrogen fertilisers can generate acidity, while a poorly managed irrigation system can produce an alkaline soil. Proper selection of crops that grow well in somewhat acidic or alkaline soils can alleviate minor soil-reaction problems. However, a trend toward high levels of either acid (pH much less than 7) or base (pH much greater than 7) in the soil is an indication of instability.

**Ss6 (Nutrient Availability: Extractable Nitrogen, Phosphorus and Potassium—N, P, K in kg/ha):** There are many laboratory tests that have been developed for measurement of 'plant available' nutrients in soil. Specific tests are designed for particular environmental regions, soil types and crops. Using a recommended test, data in units such as kilogram of nutrient per hectare (kg/ha) are obtained and the values can then be put into ranges, often defined by local soil scientists, that rate the nutrient availability on a scale of 0 to 10. The average of the scaled values for N, P and K can then be taken as a nutrient availability index.

An example of a scaling process that was developed for the study of sustainability of agricultural systems in South India measured the nutrient content of the soils by analysis for available nitrogen (permanganate extractable), phosphorus (Olsen's method) and potassium (neutral ammonium acetate extractable) in randomly sampled soils from the study villages.

Each of the parameters, measured in units of kg/ha, was then evaluated using recommendations set by other researchers in that part of India:

|          | Low    | Medium      | High   |
|----------|--------|-------------|--------|
| N (kg/ha) | <272  | 272 – 554   | >554   |
| P (kg/ha) | <12.4 | 12.4 – 22.4 | >22.4  |
| K (kg/ha) | <113  | 113 – 280   | >280   |

Scoring for the parameter is 0 for low values, 1.7 for medium and 3.3 for high.

$$\text{Nutrient availability} = Ss6s = S_N + S_P + S_K$$

where S values are the individual scores for each of the three nutrients.

**Ss7 (Saline Land Indicator):** Especially important in areas that are served by major irrigation projects is the need to prevent salinisation of the soil. For such irrigation command areas, an indicator that takes into account the portion of the land that has undergone salinisation (usually a qualitative term estimated by field observation) is required. The suggested soil salinisation indicator is:

Ss7 = Percentage of total land that in an area that is defined as saline

This state indicator can be readily scaled from 0 to 10 in the following way:

$$Ss7s = 10 - SL/3$$

where SL is the percentage land that has become saline. The denominator term is in this case an arbitrarily chosen value, defining the most undesirable situation as being when 30 per cent of the land within the command area is saline.

### Soil biological properties

Many organisms serve as bio-indicators of soil quality. A large and diverse population of microorganisms can generally be considered to be a sign of a healthy soil, degrading fresh organic matter into stable humus, releasing organic binding agents that promote good soil structure, and converting nutrients into usable forms. The many thousands of soil microorganisms play other less obvious, sometimes subtle but essential roles and a detailed assessment of their value in any particular setting is a very complex matter. Unfortunately, even basic microbiological measurements are relatively difficult to carry out in a quantitatively valid way and require specially equipped laboratories and highly trained personnel. We recommend two standard tests, one related to organic matter degradation and the other to nutrient availability. A third test, that of estimating the earthworm population, can be done easily in the field, and is a simple indication of the degree to which the soil can serve as a habitat for life.

**Ss8 (Total Soil Respiration):** As they feed off and degrade organic matter, microbes respire and in the process convert organic carbon

## Indicators for Assessing Agricultural Sustainability 143

into carbon dioxide; the rate of production of carbon dioxide can be measured in the field or on a soil sample in the laboratory. Determining the rate of production of carbon dioxide, usually expressed in terms of kilograms of carbon released to the atmosphere per hectare per day (kg C/ha/day) is a gross measure of the total population and activity of respiring microorganisms, and is therefore an indirect measure of organic matter degradation.[6] A frequently measured indicator that focuses on the *activity* of the microorganisms is to determine the respiration rate divided by the biomass carbon content. This ratio is larger for soils that have received a recent input of organic matter and it falls to a lower value as the degradation stabilises.

**Ss9 (Microbial Biomass Nitrogen):** Based on the fact that organic matter is an important source of nutrients, including nitrogen, an idea has developed that measurement of biologically active soil nitrogen (kg/ha) should be a good indicator of soil quality. There are standard (although not universally agreed-upon) laboratory methods involving fumigation that measure microbial biomass nitrogen (MBN). It is generally believed that a consistently high MBN level is indicative of good soil quality. Unfortunately, this and other methods for determining biologically active nitrogen suffer from the drawback that values depend on many extraneous factors such as the amount of moisture and organic matter in the soil. The results of the measurement are therefore dependent on climate and timing in the cropping season. For this reason, one should not rely on a single value, but look for consistency or trends and interpret the results accordingly.

**Ss10 (Earthworms):** Earthworms are considered to be a keystone species,[7] which influences and is in turn influenced by the populations of various other soil organisms. Their ability to initiate the first stages of litter degradation promotes the growth of microorganisms by providing food sources for other degraders. Earthworm mass and species richness are sensitive to short- and long-term changes in the soil environment. Because of these characteristics and because estimating numbers (expressed as mass in grams per cubic metre, $g/m^3$) in the field is possible without special training or equipment, determining the mass of earthworms can be looked upon as a basic assessment tool for soil quality. There are, however, obvious limitations to the use of this indicator. To a large degree the population of earthworms depends on the availability of moisture and fresh organic matter, which is the substrate for their energy and degradation activities. Therefore, they become in a sense an indirect measure of the amount of available organic matter in the soil rather than an intrinsic indicator of soil quality. Furthermore, their population is also influenced by factors such as compaction and chemical residues. In a very broad way then, a healthy population of earthworms is a positive sign of generally good soil quality.

> Box 3.2
> **Sustainable methods for maintaining and improving soil fertility***
>
> - Mixed livestock and crop production
> - Use of animal manure as a soil amendment, preferably after use in biogas production or composting
> - Use of farm compost, mulches and green manure
> - Recycling and composting of vegetative matter, including that from domestic sources and off-farm materials
> - No-till production methods to enhance organic matter accumulation in the soil
> - Use of intercropping, strip cropping and crop rotation
> - Incorporation of nitrogen fixing plants into rotations
> - Maintaining plants as soil cover where possible
> - Use of deep-rooting plants to recycle nutrients that have been leached to deeper levels
> - Application of physical methods such as contour bunds to minimise erosion, especially on sloping lands
>
> * Modified from Parrott and Marsden (2002).

## Water availability and stability

Water is the second component of natural capital required for productive agriculture. We emphasised previously the essential nature of water in supporting life of all kinds on our planet. This essential nature is never more evident than in the case of agriculture and all the human activities that surround it. On a world scale, 67 per cent of water withdrawals are associated with agriculture, compared with 19 per cent for industry, 9 per cent for domestic purposes and 5 per cent for other uses. In some parts of the globe, the proportion devoted to agriculture is even greater.

### Figure 3.2
Relation between resource use and productivity. In this case the resource is water. Productivity of cropping usually increases with greater availability of water but, due to problems of waterlogging and salinity, overuse of the resource has the potential to severely diminish productivity gains

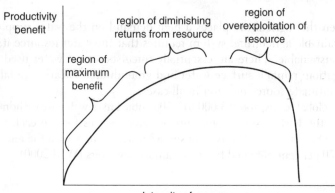

In Asia, crop production accounts for 81 per cent of freshwater withdrawals. The ratio of approximately 1,000 kg of water being required to produce 1 kg of cereal is widely quoted. The actual amount depends on the crop—some crops require even greater amounts. For rice the ratio can be greater than 3,000 kg per kg of grain. For wheat and maize, the ratio has been estimated to be between 1,400 and 1,500 kg per kg of grain in order to achieve good productivity.

It is clear therefore that large quantities of water are required to support the growing of crops, and much more to produce animal products. For example, production of 1 kg of chicken and beef require approximately 4,600 and 40,000 kg of water respectively. This is not to say that greater efficiencies cannot be achieved, but even with maximum efficiency, agriculture will remain the greatest consumer of water compared with all the other major uses around the world.

While availability of water for irrigation has the potential to improve productivity and even to improve the long-term prospects for food production, it is well-known that misuse and overuse of irrigation can operate against stability by generating salinity within the soil. This is but one example where benefits accrued using particular practices are not connected in a direct and ever-increasing linear fashion with intensity of resource use. From Figure 3.2 we can see that once the region of over-exploitation of water is reached, productivity not only ceases to increase but can dramatically, even catastrophically, fall off.

## Water quantity

Given that agriculture places a high demand on the water supply, a sustainable agricultural system requires that the water resource itself be sustainable. There are essentially three sources of water used for irrigation: rainfall, surface water and groundwater. Clearly, rainfall is the ultimate source of water in all cases.

In global terms, about 3,000 to 3,500 cubic km of water are withdrawn from the Earth's surface and groundwater reserves for use each year. Some 80 per cent of this is from surface water sources, with the remaining 20 per cent obtained from sub-surface reservoirs (WRI 2000).

### Rainfall

Rainfall patterns over time and space are largely determined by general global climatic factors. Agriculture itself is a major player, affecting precipitation patterns by its consumption and release of water to the atmosphere through respiration and transpiration of plants, and by sequestering and emitting other greenhouse gases. But it is far from the only determinant of climate and, moreover, the effects on climate are not observed on a local micro-scale. Rather, it is the cumulative and integrative effects of all human and natural processes that will determine changes in global climate.

There are, however, stability issues related to rainfall that are controllable in the local situation. One is the practice of harvesting rainfall by the building of small structures designed to minimise run-off from the local landscape. Recall the mass ratio of over 1,000 to 1 as the requirement of rainfall to produce a cereal crop. If one is aiming for productivity of 5 tonnes per hectare (10,000 m$^2$) of rainfed grain, this will require some 5,000 tonnes (5,000 cubic metres) of water. Where rainfall is the only source of water, this is equivalent to 5,000 m$^3$/10,000 m$^2$ = 0.5 m = 500 mm of rain during the growing season. Such levels of rainfall are characteristic of semi-arid regions of the world. It is important then that as much of this limited rainfall as possible be retained within the area where it falls—this is the rationale behind water harvesting. In higher rainfall areas too, such harvesting serves

## Indicators for Assessing Agricultural Sustainability 147

other useful purposes including assisting in ensuring that the local and regional groundwater resource is replenished.

### Surface water

Agriculture makes use of surface water by direct extraction from lakes, by partial diversion of flowing water in rivers, or by building dams for storing river water. In all cases, the ultimate source of the water is from rainfall. Sustainability issues therefore revert back to the way in which long-term weather patterns influence rainfall. This is clearly an issue that cannot be dealt with simply on an individual or even regional level. Other water sustainability issues centre around the lifetime of reservoirs created by dams, lifetimes usually limited by silting, which reduces the holding capacity of the reservoir.

### Groundwater

On a global scale, a growing source of water for agriculture is from groundwater reserves. Being hidden from view, there is sometimes minimal concern about the long-term effects of the massive extractions that are taking place worldwide. The lack of concern is a source of considerable problems in terms of long-term stability of high-intensity agriculture.

As we noted in describing measures for determining stability as a whole, the best way to measure stability of a water supply is to follow trends over long periods of time, but once again the time scale needed to show a trend can often be excessively long. For rainfall and for water supply from a reservoir, this is probably true, and therefore other less direct means are recommended for this purpose. It is unfortunate, however, that there are several documented cases where groundwater levels are dropping measurably over periods of time spanning a small number of decades or even years. Paradoxically, there are also cases in canal-irrigated parts of the world where the groundwater table is rising rapidly, causing waterlogging and/or a range of salinity or alkalinity problems. Therefore, an indicator for the depth of the water table is required in both these cases.

### Sw1 (Groundwater Supply):

$$Sw1s = 10 - \Delta \text{ (water table)}$$

where Δ (water table) is the change in depth (increase or decrease in metres) of water table observed over a 10-year period.

This indicator supports the idea of maintaining a stable water table. Where the water table has already been degraded (raised or lowered), restoring it to its stable 'original' value may be desirable, and the indicator can be modified accordingly. For the present version of the indicator, we select a decrease of 10 metre per decade (1 metre per year) as the poorest value, giving a score of 0, while a change of 0 metre per decade is considered optimal.

For other water quantity issues, we look at variability of the water supply, with high variability indicating instability of the system.[8]

### Sw2 (Rainfall Variability):

$$Sw2s = 10 - 10(CV - B)/R$$

where CV is the coefficient of variation for the annual total rainfall over a period of at least 10 years. Coefficient of variation is the relative standard deviation expressed in percentage terms, which is calculated as 100 times the standard deviation divided by the mean. B is the arbitrarily chosen best value that could be expected, for example 5 per cent. R is the range of possible values from the best to the poorest. For example, with an arbitrarily chosen poorest value of 60 per cent, the range is 60 − 5 = 55 per cent.

### Sw3 (Variability in Irrigation Water Availability):

$$Sw3s = 10 - (CV - B)/R$$

where CV is the coefficient of variation for the annual availability of irrigation water over a period of at least 10 years. Availability can relate to an individual field/farm, or it can be total availability in a canal serving a large command area. Coefficient of variation is expressed in percentage terms, which is calculated as 100 times the standard deviation divided by the mean. B is the arbitrarily chosen best value that could be expected. R is the range of possible values from the best to the poorest.

## Water quality

Quantity of water is therefore a major issue, but it is clear that water quality is equally important. As measures of quality, there are a large number of parameters that require consideration. General water quality parameters include:

- physical properties of temperature, colour, odour, and turbidity;
- general chemical classes of chemical properties such as pH, total dissolved solids (TSS), salinity, hardness, biological oxygen demand (BOD);
- nutrient species such as compounds containing phosphorus and nitrogen;
- specific elements including potentially toxic metals and non-metals, and organic compounds;
- radiological properties, i.e., levels of radioactivity due to particular isotopes, and;
- microbiological properties, i.e., counts of specific organisms and groups of organisms.

In deciding which criteria should be used in defining acceptable quality, end use of the water is the most important consideration. For high quality drinking water, rather stringent guidelines are in order, involving requirements within many of the classes above. Of greatest importance is a need to ensure that the water is safe with respect to its concentrations of potentially toxic chemicals as well as microorganisms. Less important from a health perspective, but still desirable, is the need to maintain aesthetic qualities in terms of colour, odour and taste. The sustainability issues related to water and human health and domestic use are considered later under the compatibility category.

For irrigation purposes, some of these quality demands are generally less severe, but other parameters are very important, including the total concentration of soluble salts, the ratio of sodium to calcium and magnesium in the water, the concentration of potentially toxic elements, especially boron, and the carbonate species concentration. Table 3.10 presents an abbreviated list of important water quality standards that apply to irrigation. Sustainability indicators are designed to take into account these quality issues.

## 150  Agricultural Sustainability

**Table 3.10**
Irrigation water quality criteria established for use in India

Salinity hazard [total dissolved solids, measured by electrical conductivity (EC in decisiemens per metre, dS/m)]

| | EC | | |
|---|---|---|---|
| | | <0.25 | excellent |
| | | 0.25–0.75 | good |
| | | 0.75–2.25 | doubtful |
| | | >2.25 | unsuitable |

Sodium hazard [measured by sodium absorption ratio (SAR) = $C_{Na}/(C_{Ca} + C_{Mg})/2)^{1/2}$]

| | SAR | <10 | excellent |
|---|---|---|---|
| | | 10–18 | good |
| | | 18–26 | doubtful |
| | | >26 | unsuitable |

Boron concentration ($C_B$ in mg/L)

| | Boron | <0.3 | excellent for all crops |
|---|---|---|---|
| | | >3.8 | very toxic for all crops |

Carbonate alkalinity (Alkalinity in milliequivalents per litre, meq/L)

| | Carbonate | <1.25 | safe |
|---|---|---|---|
| | | 1.25–2.5 | marginal |
| | | >2.5 | unsuitable |

Source: USSL (1954).

Using results from chemical analysis of irrigation water and tables of data such as these, a series of irrigation water quality indicators can be created.

### Sw4 (Water Electrical Conductivity):

$$Sw4s = 10 - (EC - 0.25) \times 5$$

where EC is the electrical conductivity measured in decisiemens per metre, dS/m.

### Sw5 (Water Sodium Hazard):

$$Sw5s = 10 - (SAR - 10)/1.6$$

where SAR is the sodium absorption ratio.

## Sw6 (Boron Toxicity Indicator):

$$Sw6s = 10 - (B - 0.3)/0.35$$

where B is the boron concentration (milligrams per litre, mg/L.)

## Sw7 (Carbonate Alkalinity Hazard):

$$Sw7s = 10 - (C - 1.25)/0.125$$

where C is the carbonate alkalinity (milliequivalent per litre, meq/L).

In each of the above four water quality indicators, the calculated values for very high quality water could be greater than 10. In these cases, a value of 10 should be assigned.

All of the indicators suggested here in the stability category are either pressure or state indicators. An indicator such as the Electrical Conductivity indicator describes a pressure on the soil water system. In the short term it may not affect productivity, but continuing poor values are predictors of future problems. Others such as the Soil Nutrient Availability indicator describe the quality of the present state of the soil. Response indicators can also be created for specific situations. These describe activities that serve to enhance the quality of the basic resources for crop production.

Depending on individual situations, many soil and water conservation activities are practiced; these can be subjects around which response indicators are created as in the following examples:

- Windbreaks to minimise erosion by wind: measured as length of windbreaks per hectare of arable land
- Water harvesting strategies: fraction of farmers using practices to trap soil moisture
- Land improvement by cropping strategies: for example, an indicator can be based on the ratio of high residue crops or soil-building crops such as alfalfa or clover to total land area
- Land improvement by modified tillage methods: ratio of area under conservation tillage to total land area
- Methods to prevent salinisation: percentage of farms that have installed acceptable drainage systems to minimise waterlogging and salinisation or alternatively used other methods such as planting appropriate tree species for salinity control.

- Extent of reuse of irrigation water, as percentage of drainage water, where quality is acceptable and can be maintained
- Cropping methods to cope with saline soils: percentage of saline land occupied by salt tolerant crops, preferably fodder crops, or those useful for other forms of biomass production, such as fuelwood
- Reclamation measures: ratio of area of land being subjected to reclamation strategies, such as application of gypsum and provision of drainage to reclaim salt affected areas

---

**Box 3.3**
**Strategy for assessing stability**

- The scope of the assessment should be determined (area and time to be covered).
- Determine as many indicators as possible from those suggested for measuring soil quantity (Ss1), soil quality (Ss2 to Ss10), water quantity (Sw1 to Sw3), and water quality (Sw4 to Sw7). Some of the indicators require development of quantitative procedures that will apply in each specific situation.
- Scale each of the chosen indicators to a common scale.
- Calculate the average within the four sub-category values, giving a measure of quantity and quality for both soil and water.
- The four sub-indices can then be averaged to produce a single stability index.
- Where appropriate, develop and evaluate response indicators for stability in order to assess measures that are being undertaken to improve the quantity and quality of the crop production resources.
- Consider how the issues related to durability are connected with those of stability in both a positive and a negative way. Some of the response indicators that are appropriate for durability also apply in the case of stability.

---

In the processes described here for indirect measurement of stability of the agricultural system, a large number of individual indicators may have been determined, depending on availability of data. The benefits

Indicators for Assessing Agricultural Sustainability 153

and problems of aggregating results into a single index have been discussed in Chapter 2 and are illustrated by this exercise. Following the strategy for assessing stability, the indicators are combined by simple averaging to give a single overall index value that provides an estimate of potential stability or instability of the system.

If a clearly high or low overall value is obtained, there is an obvious conclusion that one could predict prospects for continuing productivity (high value) or declining productivity (low value) over time. On the other hand, it will frequently be found that the overall stability index falls in a middle range. This type of value has very limited significance. It could be made up of a large number of mostly middle-range results or, alternatively, it could indicate that certain stability features are favourable while others fall in a distinctly unfavourable range. In this situation it is clearly important to disaggregate the combined result in order to examine the individual components. The disaggregation might reveal a problem in one of the four areas. Further disaggregation of the individual indicators used to describe that aspect of stability might then pinpoint specific problems and virtues.

## 3.3 Efficiency

*Efficiency* To be sustainable, all the resources required for agriculture—human, animal and material—should be used in a way that is not wasteful, but maximises output per unit input.

The processes of crop production require a range of inputs and efficiency is the measure of the extent to which those inputs enhance the crop yield. An efficient operation is one where a small level of inputs results in excellent productivity. Inefficient operations are wasteful of resources or other inputs, and are in this respect not sustainable.

Inputs of many kinds work together in the entire process of producing crops, and they include components from all the types of capital available:

- Natural capital in the form of soil, water and sunlight
- Human capital such as agricultural research, planning or physical labour

- Social capital by way of extension services, government support, etc.
- Physical capital in the form of machinery and other material input (seeds, agrochemicals, manures)
- Financial capital to purchase the materials required for carrying out production

Likewise, outputs can fall into a similar range of categories

- Natural capital produced (or lost) in the form of altered landscapes, and in the way in which habitats for other species are created or destroyed
- Human capital generated by the new knowledge gained through participation in the farming process
- Social capital enhanced by sharing the new knowledge as well as by sharing in the physical and other operations that go into agricultural production and distribution of produce
- Physical capital represented by the crop itself, along with by-products that can also serve utilitarian functions
- Financial capital gained when surplus crop over the farmer's personal needs is produced and sold

Efficiency involving any combination of these categories is important. For example, scientific research applied to agriculture (as an input) should lead to greater crop yields, or improved production methods or benefits such as a reduced requirement for agricultural chemicals (as outputs). In the case of research, it is seen that many of the 'inputs' involved are intangible and usually difficult to express in quantifiable terms. That does not mean, however, that the intangibles should be neglected when attempting to evaluate sustainability. This can be done by providing a descriptive picture of the impact that the research and extension activities have had on supporting practices that lead to stability.

In contrast to the intangibles, there are physical aspects of agriculture that can be more readily described by using quantitative means, and it is for these aspects that we can determine efficiency and create indicators in a relatively simple way.

Figure 3.3 depicts the general agricultural process, showing the variety of natural and anthropogenic physical inputs that go into crop production, as well as the range of products that result from this process.

## Indicators for Assessing Agricultural Sustainability

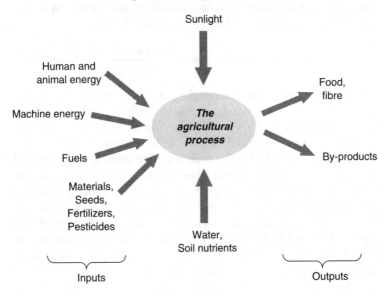

Figure 3.3
The agricultural process involves a variety of physical inputs, all of which can be described in both energy and monetary terms. Likewise, the products, both primary and secondary, have energy and monetary values. In this diagram, sunlight, water and soil nutrients are considered as natural 'givens', although it is recognised that in irrigated systems water will also have quantifiable energy and financial values

Efficiency in a general sense can be measured as a ratio of output (crop yield, or yield of some portion of the crop, expressed in physical units of various types or in monetary value) over input of the particular type selected. One basic efficiency relationship is:

$$\text{Agricultural efficiency} = \frac{\text{Output (crop)}}{\text{Inputs (various)}}$$

In this calculation, any one of the inputs required for plant growth could be used individually or in combination. For example, the input could be human labour or fuel cost or non-renewable resource use. Likewise, output can be measured using various monetary or physical terms.

Another related equation considers efficiency in output gain per unit input:

$$\text{Agricultural efficiency} = \frac{\text{Output} - \text{Input}}{\text{Input}}$$

In this equation, both output and input must be expressed in the same units, typically energy or monetary units.

An alternative expression by which efficiency can be measured also makes use of the difference between output and input, but is normalised to a specific area of land, typically one hectare:

$$\text{Agricultural efficiency} = \frac{\text{Output} - \text{Input}}{\text{Area}}$$

There is a vast array of possibilities, from which we shall choose a small number. In the first instance, let us consider output. We have seen that there is a primary and a secondary product for most crops, the two together making up the total biomass. In the case of wheat, for example, the primary product is the grain itself but straw is also produced. On a mass basis, the amount of straw is usually similar to the amount of grain. The exact mass proportion of each depends on the variety, as well as on specific environmental and management conditions. The straw (or other secondary product) is sometimes considered waste and may be destroyed by burning or other means. But it can also be considered a resource, and this is especially important if the goal is a sustainable agricultural system. Almost every secondary product of food and fibre production can be used in a variety of ways: as a soil amendment, as cattle fodder, as a source for biogas production or to be burned directly for cooking or heating purposes. Therefore, one decision that is required in estimating efficiency is whether to use the secondary product in the output component of the calculation.

With regard to inputs, there are even more options to be considered. Measuring efficiency in terms of land area, e.g., the amount of land required to produce one tonne of corn, would be one obvious possibility. This measure, however, is just the inverse of productivity—area per output rather than output per area—and is therefore not a indicator recommended as an additional efficiency measure. The other input parameters like human labour, total resources or non-renewable resources are more appropriate inputs for determining efficiency. One way of quantifying these resources would be in financial terms. In that case, it would be logical to put a monetary value on output as well. Efficiency would then take the form of an economic parameter.

If, however, the plan is to measure efficiency in terms of the physical components of the agricultural system, we need some kind of 'common currency' that can be used to measure all the different aspects of agricultural production. It is with this in mind that we will describe a rational way in which physical efficiency can be evaluated.

In a very fundamental sense, agriculture depends primarily on the natural capital available to the system, on the soil qualities, the availability of water, and the intensity and duration of sunlight. It is for this reason that maintenance of the integrity of the natural capital is the subject of stability indicators. The natural processes that make use of this capital are many, but one process is particularly fundamental. This is the ability of plants to convert sunlight into plant material (biomass, principally carbohydrate), a process that is described by the simplified equation for photosynthesis (see also Chapter 1):

$$\text{Carbon dioxide} + \text{water} \xrightarrow{\text{sunlight}} \text{organic matter} + \text{oxygen}$$

$$CO_2 + H_2O \xrightarrow{h\nu} \{CH_2O\} + O_2$$

Assuming that carbon dioxide, water and plant nutrients are available, photosynthesis will occur in the green portions of plants, enabling them to grow, with a resulting production of organic biomass. Knowing the total number of hours of sunlight and its intensity, we can calculate the efficiency of conversion of *solar energy* into *energy embodied in the plant material*. In other words, photosynthesis is a process by which one form of energy is changed into a different form. Even under optimum conditions, for almost all plants the conversion efficiency is less than 1 per cent, with only occasional exceptions occurring in the case of highly productive, fast-growing species.[9] Sugarcane is one of the best examples of a plant with large solar energy efficiency. For sugarcane, efficiencies of 3–4 per cent have been reported, leading to very high biomass productivity (see Tables 3.1 and 3.2). Ultimately, it will be the photosynthetic efficiency that is the limiting factor in plant growth; for this reason increasing the intrinsic ability of plants to use sunlight in the production of biomass is the subject of ongoing research. For most farming situations, however, it is not particularly helpful to calculate solar conversion efficiency as there are usually other factors that limit a particular plant's ability to achieve optimum photosynthetic efficiency. Some of these factors include the availability of water, soil physical properties and nutrients. In order to maximise production, humans are involved in making decisions that optimise the use of the natural resources. Humans also make choices about the use of other human-made inputs that have the potential to enhance production even further. In a practical sense, these other inputs are the true factors that limit plant growth.

## Water use efficiency

Water is one of the essential natural factors. As we have seen, in the case of many crops upwards of 1,000 kg of water are required to produce 1 kg of grain (or 2 kg of biomass). In rainfed areas, it might be assumed that humans have little control over water availability. In actuality, however, roof water collection systems, contour strips, diversion structures, ground water recharge systems, run-off farming, bridges that serve as barrages or large bunds, and dams are some of the common methods used in harvesting rain water.

Because of the essential nature of water and because of its scarcity in many areas, a water use efficiency indicator can be applied.

### Efw1 (Water Use Efficiency):

$$Efw1s = (P/W) \times F$$

where P is the product output in tonnes per hectare and W is the water used in tonnes per hectare. F is defined below.

The amount of water supplied can be calculated from the rainfall that falls during the growing period. For example for 500 mm (0.5 m) of rainfall falling in a 1 ha (10,000 m$^2$) area, the volume of water is $0.5 \times 10,000 = 5,000$ m$^3$ (cubic metres). Because a cubic metre of water weighs one tonne, the mass of water is 5,000 tonnes.

F is a factor determined for each situation. For example, if it is assumed that optimal water use efficiency for a particular crop is 1,000 tonnes per 1 tonne output, then a F value of 10,000 is selected. This scales the results within our usual 0 to 10 scale. As an example, take a situation in an arid region where sorghum is raised during a season when there is 450 mm of rain. The output is 2.3 tonnes of grain per hectare. Assume that the optimal efficiency is represented by production of 1 tonne of sorghum grain for every 1,000 tonnes of water. The value of Ew1s is then

Efw1s = (P/W) × 10,000 = (2.3/4,500) × 10,000 = 5.1

In this example, only the grain portion of the crop was used in calculating the output. If desired, depending on the crop and the specifics of each situation, the calculation (including F value) can be modified to include both the primary and secondary products.

## Energy as an integrative measure of efficiency

Every physical component of agricultural production, both inputs and outputs, can be expressed in energy terms. For calculating outputs, data concerning the calorific content of foods are widely available. Less readily available are the energy content values for materials such as straw, cotton sticks and seeds, groundnut shells, and most other secondary products. For inputs, energy values have been measured or estimated for human and animal labour, fuels for machinery, manures, synthetic fertilisers, pesticides and the embodied energy of hand implements and machinery. The energy content of fossil fuels as well as of biomass (including animal manures) is measured by determining the heat output during combustion. Energy for animate labour is calculated from extensive studies determining food requirements required to sustain the human being or the animal over time when carrying out specific types of labour. Calculation of the energy embodied in various manufactured products is a complex process involving estimates of energy required to extract, refine, synthesise, assemble and transport all the materials that go into the manufacture of a particular product. For products that are not consumed but used over a period of time (e.g., machinery), the amount of energy required is then averaged over the assumed lifetime of the product. Especially in this calculation, there is considerable uncertainty, but this is the case in general for all energy estimates. Depending on the source of information, there may be a large range of values given for the same factor. Table 3.11 reports estimates of energy in all the categories described here. In the table, the unit of energy chosen is the recommended SI unit, the joule (J). Examples of calculations using these data are given later.

Having a complete selection of energy values, we can carry out a series of integrative efficiency calculations. The general form of many of these calculations is:

## Table 3.11
Energy equivalents associated with agricultural activities—chemicals, manufactured materials, human and animal labour, and products

| Component | Energy in J | per | Source[a] | Rationale |
|---|---|---|---|---|
| CROPS—PRIMARY AND SECONDARY | | | | |
| Seed | | | | |
| General, non-hybrid | $1.5 \times 10^7$ | kg | A | Based on data for average energy value of product crops |
| Paddy, hybrid | $1.7 \times 10^7$ | kg | A | |
| Sorghum, hybrid | $5.9 \times 10^7$ | kg | A | |
| Cotton | $4.4 \times 10^7$ | kg | A | |
| Groundnut, hybrid | $7.6 \times 10^7$ | kg | A | |
| Oilseeds & legumes | $3.5 \times 10^7$ | kg | A | Average for various oilseeds and legumes in reference A |
| Paddy | $1.23 \times 10^7$ | kg | A | |
| Sorghum/Millets | $1.37 \times 10^7$ | kg | A | |
| Legumes | $1.42 \times 10^7$ | kg | A | Based on data for peas |
| Cowpea | $1.44 \times 10^7$ | kg | E | |
| Groundnut | $1.63 \times 10^7$ | kg | E | |
| Oilseeds | $2.5 \times 10^7$ | kg | A | |
| Corn | $1.45 \times 10^7$ | kg | A | |
| Tomato | $8.4 \times 10^5$ | kg | A | Based on values for other vegetables, by wet weight |
| Chili | $7.5 \times 10^5$ | kg | B | By field weight |
| Banana | $2.9 \times 10^6$ | kg | A | |
| Mango | $2.8 \times 10^6$ | kg | A | Assumed to be similar to peach: value from DP |
| Sugarcane | $1.42 \times 10^7$ | kg | B | 7.5% juice at $1.6 \times 10^7$, cane at $1.4 \times 10^7$ |
| Watermelon | $5.03 \times 10^5$ | kg | A | 1,832,760 kcal in a harvest of 15,273 kg in India |
| Cotton seed | $2.3 \times 10^7$ | kg | G | Assume crop is 17% seed |
| Cotton fibre | $1.8 \times 10^7$ | kg | G | Assume crop is 9% fibre |

| | | | | |
|---|---|---|---|---|
| Cotton sticks | $1.7 \times 10^7$ | kg | G | Assume crop is 74% sticks |
| Straw in general | $1.4 \times 10^7$ | kg | | Generic value for 'waste' agricultural products |
| Green manure | | | | |
| Hay, mixed | $7.2 \times 10^6$ | kg | A | Based on field dried weight |
| Alfalfa | $8.9 \times 10^6$ | kg | A | Based on field dried weight—15% moisture |
| **FERTILISERS** | | | | |
| Nitrogen fertiliser (general) | $6 \times 10^7$ | kg | J | per kg N |
| Nitrogen fertiliser (anhydrous ammonia) | $5 \times 10^7$ | kg | A | per kg N |
| Nitrogen fertiliser (ammonium nitrate) | $6.16 \times 10^7$ | kg | A | per kg N |
| Nitrogen fertiliser (other nitrate) | $7.8 \times 10^7$ | kg | A | per kg N |
| Nitrogen fertiliser (ammonium sulphate) | $6.6 \times 10^7$ | kg | L | per kg N |
| Nitrogen fertiliser (Urea) | $6 \times 10^7$ | kg | A | per kg N |
| Phosphate fertiliser (rock phosphate) | $5.45 \times 10^6$ | kg | A | per kg $P_2O_5$ |
| Phosphate fertiliser (single superphosphate) | $9.6 \times 10^6$ | kg | | per kg $P_2O_5$ |
| Phosphate fertiliser (triple superphosphate) | $1.25 \times 10^7$ | kg | A | per kg $P_2O_5$ |
| Potash | $6.7 \times 10^6$ | kg | A | per kg $K_2O$ |
| | $4.2 \times 10^6$ | kg | B | per kg KCl (recalculated from above value) |

*(Continued)*

*(Continued)*

| Component | Energy in J | per | Source[a] | Rationale |
|---|---|---|---|---|
| Gypsum | $1 \times 10^7$ | kg | A | |
| DAP | $6 \times 10^7$ | kg | A | |
| 17-17-17 | $1 \times 10^7$ | kg | A | Using values from A, based on 17% ammonium, 17% phosphate rock, 17% potash |
| 19-19-19 | $1.1 \times 10^7$ | kg | A | Using values from A, based on 19% ammonium, 19% phosphate rock, 19% potash |
| 20-20-20 | $1.2 \times 10^7$ | kg | A | Using values from A, based on 20% ammonium, 20% phosphate rock, 20% potash |
| Micronutrients | $1 \times 10^8$ | kg | A | |
| FYM | $3 \times 10^6$ | kg | A | Fresh (78% moisture content: from reference F) |
| | $1.25 \times 10^7$ | kg | A | Dry, assumes 80% moisture |
| Sheep/goat manure | $6.57 \times 10^6$ | animal-day | A, K | Per animal day, uses deposition of 2.19 kg, and assumes energy equivalent to that of fresh FYM |
| **PESTICIDES** | | | | |
| Liquid pesticide (generic) | $3.63 \times 10^8$ | L | A | |
| Solid pesticide (generic) | $3.11 \times 10^8$ | kg | A | |
| Carbofuran | $4.54 \times 10^8$ | kg | L | Specific value for particular surveys |
| Phorate | $1.72 \times 10^8$ | kg | L | Specific value for particular surveys |
| Roundup | $4.54 \times 10^8$ | kg | L | Specific value for particular surveys |

## HUMAN AND ANIMAL LABOUR

| | | | | |
|---|---|---|---|---|
| Human labour (general) | $1.6 \times 10^6$ | hour | E | $1.4 \times 10^6$ is equivalent to walking carrying a heavy load |
| Human labour (watch and ward, irrigation) | $1.1 \times 10^6$ | hour | | |
| Human labour (harvesting, threshing) | $2.0 \times 10^6$ | hour | E | Slightly less than equivalent for energy spent cutting wood |
| Bullock pair + one person | $2.5 \times 10^7$ | hour | E | Pimentel (1980) reported that a bullock pair (during tillering) expended 260,000 kcal in 65 h; this value reflects a bullock pair and a driver |

## MACHINERY AND FUELS

| | | | | |
|---|---|---|---|---|
| Machinery (metal) | $2.4 \times 10^4$ | kg-hour | | Uses production energy of 0.4 GJ/kg, with a lifetime of 15 yrs and average use of 3 hrs per day. Calculated per kg of machine, per hour |
| | $3.8 \times 10^3$ | kg-hour | A, H | Uses production energy of 0.01 GJ/kg, materials energy of 0.06 GJ/kg, with a lifetime of 15 yrs, and average use of 3 hrs per day (1976 data). Calculated per kg of machine, per hour |
| | $5 \times 10^3$ | kg-hour | H, J | Uses production energy of 87 MJ/kg, with a lifetime of 15 yrs, average use of 3 hrs per day. Calculated per kg of machine, per hour |
| | $4.5 \times 10^3$ | kg-hour | L | Uses lifetime of 15 years, average use of 3 hrs per day, and referenced values for intrinsic and production energy (1968 data, taken from reference M). Calculated per kg of machine, per hour |

*(Continued)*

*(Continued)*

| Component | Energy in J | per | Source* | Rationale |
|---|---|---|---|---|
| | $4.3 \times 10^3$ | kg-hour | M | Based on a literature value for production energy for agricultural implements of 70 MJ/kg and assumptions as above (1968 data). Calculated per kg of machine, per hour |
| Machinery (wood) | $1.2 \times 10^4$ | kg-hour | | |
| Diesel fuel | $4.8 \times 10^7$ | L | A | Inherent energy plus energy of production |
| Machinery (fuel for ploughing) | $3.36 \times 10^8$ | hour | H | Assumes consumption of 7 L of diesel per hour |
| Machinery (fuel for trolleying) | $2.16 \times 10^8$ | hour | H | Assumes consumption of 4.5 L of diesel per hour |
| Machinery (fuel for tillering) | $2.88 \times 10^8$ | hour | H | Assumes consumption of 6 L of diesel per hour |
| Machinery (fuel for harrow) | $3.12 \times 10^8$ | hour | H | Assumes consumption of 6.5 L of diesel per hour |

**Sources:**

A. Pimentel (1980).
B. Gupta et al. (1984).
C. Rao, A.R. (1985).
D. Aggarwal, G.C. (1989).
E. Pimentel and Pimentel (1979).
F. Dahiya and Vasudevan (1986).
G. Rao et al. (1992).
H. Verma and Pathak (1998).
I. Shyam and Gite (1990).
J. Loomis and Connor (1992).
K. Prasad (1995).
L. Fluck and Baird (1980).
M. Makhijani and Lichtenberg (1972).

* Values without sources are estimates by the authors.

$$\text{efficiency} = \frac{\text{energy content of the crop}}{\text{energy value of inputs}}$$

Using this and related general formulas, a large range of energy-based efficiency indicators can be envisaged. Since the calculation is a ratio of energy values (with energy expressed in joules), the ratio has no units. Note that the ratio can have a value not only of less than 1 but also exceeding 1, depending on what inputs and outputs are used in the calculation. To convert the ratio into an indicator, a rating scale is established, depending on the particular situation. One measure (Heinonen 2001) of overall efficiency defines the system as being poor for ratios less than 1, and excellent when the ratio is more than 10. However, there are many issues that go into decisions regarding the rating scale. Some suggested energy-based indicators are given here.

**Efe1 (Primary Energy Efficiency):** This index uses the value of primary energy efficiency: the energy ratio of output in the form of principal component of the crop (e.g., grain) over the sum of all inputs used to produce the crop.

$\text{Efe1}$ = primary crop output energy/energy content of all inputs

$\text{Efe1s}$, primary energy efficiency (scaled) is determined as above and given a scaled value using a scoring system such as the one given below. This scale is given as an example and should not be considered to be applicable in every situation.

| Ratio | Indicator value |
|---|---|
| >8 | 10 |
| 7–8 | 9 |
| 6–7 | 8 |
| 5–6 | 7 |
| 4–5 | 6 |
| 3–4 | 5 |
| 2–3 | 4 |
| 1–2 | 3 |
| 0.5–1 | 2 |
| 0.25–0.5 | 1 |

There are other energy-based efficiency indicators. Some important and useful examples follow.

**Efe2 (Total Energy Efficiency):**

Efe2 = energy content of all products/energy content of all inputs

A scaled indicator value (Efe2s) is determined from an appropriately designed table of the type suggested above.

**Efe3 (Non-Renewable Primary Energy Efficiency):** In this indicator, renewable energy includes human and animal energy, plant and animal manures, and seeds produced on the farm. Non-renewable energy is that derived from use of machines, fuels, synthetic fertilisers and pesticides, and commercially produced seed stocks.

Efe3 = Energy content of primary product/Non-renewable energy inputs

A scaled indicator value (Efe3s) is determined from an appropriately designed table of the type suggested above.

**Efe4 (Renewable Energy Use):** Besides measuring output/input ratios, there are other means in which the efficiency of resource use can be measured. For example, a Renewable Energy Use indicator can be estimated. Renewable inputs include plant and animal manures and animal power for field operations, while non-renewable inputs are chemical fertilisers and pesticides, and machinery whose power is supplied by fossil fuels. This indicator assumes that renewable energy source consumption is more sustainable than consumption of non-renewable sources. It also recognises that some non-renewable energy use is inevitable. In this instance, optimal use of renewable energy is taken to be 70 per cent of total energy use, while the least satisfactory situation is where 0 per cent of the inputs are from renewable sources. To scale the indicator in the usual way, the indicator is calculated as:

$$\text{Efe4s} = \% \text{ renewable energy}/7$$

Therefore, if 40 per cent of the input energy can be categorised as renewable, the value of the indicator is 40/7 = 5.7. It is worth noting here that there will be circumstances in which more than 70 per cent of the energy comes from renewable sources. In these cases, the calculated indicator value will be greater than 10, in other words higher than 10 on the defined scale. It is recommended that, though this may suggest better than optimal sustainability conditions, the value still be reported as 10.

## Energy efficiency calculations

As an example of the energy calculation methodology we consider a situation where an improved variety of sorghum was grown on an eight-hectare field in Karnataka State in India. The materials that were employed for crop production included seed and urea fertiliser. A tractor was used for initial ploughing, and also to power a threshing unit. Harrowing, sowing and intercultivation were done using a bullock pair as the source of power. These and other operations—fertiliser application, weeding, harvesting, packing—also required human labour. Data for all these processes are summarised in Table 3.12.

## Nutrient-based efficiency indicators

Having emphasised the value of using energy as an integrative common currency for efficiency calculations, we can also note that other efficiency measurements are also appropriate for specific purposes. Besides sunlight and water, nutrients are another essential requirement for plant growth. Many of these are also outputs that are essential for human nutrition. There are approximately 16 elements that are considered to be essential and any one of these could be a limiting factor to good productivity in particular situations. Nitrogen, phosphorus and potassium are in most cases the nutrients required in the largest amounts and we will restrict our selection of indicators for nutrient use efficiency to these three elements. Using the standard 'output over input' formula again, the nutrient use efficiency in the case of nitrogen would be expressed as:

$$\text{Nitrogen use efficiency} = \frac{\text{Nitrogen removed by crop}}{\text{Nitrogen inputs from all sources}}$$

**Table 3.12**
Sample calculation for some aspects of energy efficiency

| | | Energy content | Total energy ($10^9$ joules, GJ) |
|---|---|---|---|
| *Output parameters* | | | |
| Total grain harvest | 8,500 kg | $1.37 \times 10^7$ J/kg | 117 |
| Estimated mass of stalks | 10,600 kg (estimate 1.25 kg stalks per 1 kg grain) | $1.4 \times 10^7$ J/kg | 150 |
| Total production | | | 270 |
| *Input parameters* | | | |
| Materials | | | |
| Seed | 65 kg | $5.9 \times 10^7$ J/kg | 3.8 |
| Urea | 500 kg | $2.8 \times 10^7$ J/kg | 14* |
| Human labour | 171 person days | $1.2 \times 10^7$ J/day (est. 6 hour day) | 2.1 |
| Bullock labour | 32 bullock pair days | $1.5 \times 10^8$ J/day (est. 6 hour day) | 4.8 |
| Machine | 10 hours (tractor) | Embodied $7.7 \times 10^6$ J/hour ($2.4 \times 10^4$ J/kg/hour $\times$ 500 kg $\times$ 10 h) | 0.12* |
| | 7 hours (thresher) | $1.1 \times 10^7$ J/hour ($2.4 \times 10^4$ J/kg/hour $\times$ 600 kg $\times$ 7 h) | 0.10* |

|  | Fuel |
|---|---|
| Total energy | $3.4 \times 10^8$ J/hour  5.7* |
| Total non-renewable energy | $(4.8 \times 10^7$ J/litre $\times 7$ litres/hour $\times 17$ h)  31  20 |

* aspects of the operations that consume non-renewable energy

Using these data, various efficiency calculations (output/input) include:

Overall energy efficiency = 270/31 = 8.7 (scaled value = 5)
Primary energy efficiency = 117/31 = 3.8
Non-renewable primary energy efficiency = 117/20 = 5.9 (scaled value = 5)
Renewable energy fraction = 35%

In this instance, the amount of nitrogen in the crop and the amount added as animal manure and synthetic fertilisers is usually easy to estimate, but there is difficulty in estimating how much nutrient is obtained from soil. For this reason, a more readily calculated modified parameter might measure the *added nitrogen use efficiency*, where the only input considered is the nitrogen provided as a soil amendment. Efficiency calculations for other nutrients can be worked out in the same way.

### Efn1 (Added Nitrogen Efficiency):

$$Efn1 = \text{nitrogen content of crop/added nitrogen}$$

Both the nitrogen content of the crop and the nitrogen added would be calculated in the same mass units, typically kilograms. Again there is a question as to how this should be scaled. A ratio of 1 indicates that the same amount of nitrogen that was added to the soil was assimilated by the crop, a value greater than one suggests that soil nitrogen was 'mined', while a value less than one indicates that some of the added nitrogen was not used by the plant and perhaps lost from the surrounding soil—in many cases leached into surface or groundwater or released into the atmosphere by processes such as denitrification. The ideal then is a value near 1 or slightly greater than 1. A slightly greater value does not suggest that the nitrogen content of the soil would be in danger of eventually decreasing to zero, since this element is supplied through natural renewal processes, including available forms being carried in by rain and produced through nitrogen-fixation reactions of soil organisms. A scaling table to give values for En1s takes the following form:

| Ratio | Indicator value |
|---|---|
| 1–1.1 | 10 |
| 1.1–1.2 or 1–0.95 | 9 |
| 1.2–1.3 or 0.95–0.9 | 8 |
| 1.3–1.4 or 0.9–0.85 | 7 |
| 1.4–1.5 or 0.85–0.8 | 6 |
| 1.5–1.6 or 0.8–0.75 | 5 |
| 1.6–1.7 or 0.75–0.7 | 4 |
| 1.7–1.8 or 0.7–0.65 | 3 |
| 1.8–1.9 or 0.65–0.6 | 2 |
| 1.9–2.0 or 0.6–0.55 | 1 |
| >2.0 or <0.5 | 0 |

This table is not applicable to low-productivity crops grown with little or no added nutrient. In these cases, most or all of the nitrogen is derived from the soil and nutrient use ratios could be much higher than 2. Yet, if the productivity is small, the amount of nitrogen extracted could still be within the sustainable carrying capacity of the soil.

Other nutrient ratios are calculated and scaled in a similar way. An important difference is that for these elements there are no significant renewal mechanisms except those provided by the addition of renewable (manures, etc.) or non-renewable (chemical fertilisers) sources of the nutrients.

### Efn2 (Added Phosphorus Efficiency):

$Efn2$ = phosphorus content of crop/added phosphorus

### Efn3 (Added Potassium Efficiency):

$Efn3$ = potassium content of crop/added potassium

## What should be done with secondary biomass?

One of the central features of any discussion on agricultural sustainability must be the question, 'What is the role of secondary biomass in a sustainable agricultural system?' This question is especially relevant in examining issues of efficiency, since a decision must be made regarding whether to incorporate the secondary material into calculations. Whether or not secondary material should be accounted for in production and efficiency estimates depends on the nature of the material and the uses to which it can be put.

As a general term, biomass refers to the carbon-containing matter produced in nature by biological processes involving microorganisms or macroorganisms. We return here to the broad issues of ecology. In any natural ecosystem, plants, animals and microorganisms are all components of complex networks through which energy and chemicals flow. In a highly developed 'wild' system, plants and other autotrophic organisms function as producers of organic matter through photosynthesis and other pathways. Animals and other heterotrophic organisms consume the primary material, in the process converting it into other forms of biomass: structural material, specific molecules that provide specialised metabolic functions, and wastes. Together, the various components of living matter are able to first capture solar energy, synthesising it into 'useful' products, and then to conserve, transform and utilise it in many different ways—all contributing to the diversity we see around us.

The ultimate driving force for the production of biomass is solar energy, which makes photosynthesis possible. As stated earlier, photosynthesis is intrinsically inefficient in that most of the energy reaching the Earth is not used to produce living matter. In the first place, only a small portion (the visible green light) of solar radiation is actually usable by the plant. Then, about a third of this usable energy goes into respiration, a non-productive process of the growing plant. Typically, considerably less than 1 per cent of the total solar energy reaching the Earth's surface is actually involved in converting carbon dioxide and water into biomass. Of this small fraction, for crop plants about 45 per cent goes into building the structural material—roots, stems and leaves—and another 45 per cent supports reproduction through the production of seeds. The remaining approximately 10 per cent is lost due to animals feeding off the living, growing plants.

In terms of human needs, the principal nutritive and economic value of most field crops is found in the seeds. This is what we call the primary product. The remaining, inedible structural material, the secondary product, could be, and sometimes is, considered 'waste'. The secondary material is, however, potentially a highly valuable resource, and in any sustainable system, careful planning should go into making sound decisions about its fate.

Besides field crops, there are other forms of biomass that are associated with diverse agricultural systems. One form is wood derived from trees growing within the area set aside for crop growth. There are several reasons why promoting tree growth as a component of an agroecosystem is an important activity. Compared to field crops, trees are more efficient at converting solar energy into biomass, with productivities ranging from 12 tonnes (in temperate forests) to 18 tonnes (in highly productive tropical forests) total mass per hectare per year. As a means of sequestering atmospheric carbon dioxide, this is a highly desirable practice in terms of minimising greenhouse gas accumulation in the atmosphere with its long-term effects on global climate. Added to this are other benefits from growing trees within an agroecosystem: provision of diverse habitats for fauna, control of wind and water erosion, water holding ability, biological control of salinity, and filtering of dust particles from the atmosphere, to name a few.

Among living things, at the other end of the size scale from trees are microorganisms. While not visible or obvious, microorganisms in the soil can also contribute significantly to the total mass of biological material in a given area. For well-maintained soils, the mass of bacteria, fungi and other microorganisms can be up to 2 tonnes per hectare. Where soil fumigation is used, this number is reduced to near zero.

Microorganisms serve many useful functions, especially ones that are important in terms of nutrient cycling. For example, it is through bacteria and other microbes that dead roots from previous crops are decomposed. The nutrients like nitrogen are then released as inorganic forms by a process called mineralisation, and then transformed through the assistance of soil microorganisms into products such as ammonium and nitrates that serve to support plant growth. It was due to this that maintenance of a healthy microbial population was earlier described as a stability issue.

Biomass is also derived from animals. Animals in the soil such as earthworms serve important functions in converting raw organic matter into more readily decomposable forms. They also contribute to aeration of the soil, enhancing water movement and minimising problems of waterlogging. Where domestic animals are raised in a mixed farming system, they also play a role in cycling the organic matter within the system. While they consume plant material, they also convert a substantial portion of it into urine and faeces. Average daily output from domesticated farm animals is around 6 per cent of the animal weight. A 500 kg animal, for example, produces about 30 kg of manure every day. Where there is a large herd of cattle, or herds of sheep or goats, or a hog or chicken farm in which the animals are kept in an enclosed area, vast amounts of animal waste are generated. This must be disposed of (to be more accurate, 'used') in a sustainable and productive fashion.[10]

What is the best use of these various secondary materials from agricultural production? There are three principal options for use of residual plant material and animal wastes:

- As animal fodder (applies to plant materials only)
- As a soil amendment to build up the organic matter content and as a source of macro- and micro-nutrients
- As a fuel, either to be burned directly, or after conversion into some other more convenient solid, liquid or gaseous form

There are advantages and disadvantages associated with each of these options.

### Biomass as animal fodder

Some secondary material from field crops is highly valuable as animal fodder. The value is so great that, in particular instances, it can be the determining factor in making choices regarding crops. In the arid Deccan Plateau region of south-central India, sorghum is widely grown. Many farmers have continued to grow traditional or improved varieties that head out on a tall (2 m or more) stalk, in preference to the higher-yielding (in terms of grain) short-stemmed hybrids. This is because the stalks are a very desirable fodder source for bullocks and milk cattle. Secondary material from other crops is also a rich source of nutrients. A particularly good example of high-nutrient material is the residual biomass of the groundnut plant after removing the underground seeds. On the other hand, some secondary products are nutrient-poor and are less widely used as good animal fodder. Rice straw is a good example of this. Obviously, wood and branches from trees are also inedible, and some other option must be chosen for their use. Assuming good nutrient content and digestibility then, a first choice for use of the secondary product could be as animal food. The other options would be considered only as a lower-level option.

### Biomass as soil amendment

It is well documented that organic matter serves a number of useful functions in the soil. Concerning physical properties, it increases water holding capacity in sandy soils and reducing the possibility of premature drying under low rainfall conditions. With heavy clayey soils, it breaks up the blocky structure, helping to form smaller aggregates that allow for free water movement, thus reducing the possibilities of waterlogging. Organic residues of crops also contain small but significant amounts of nutrients. Typical nutrient contents of stalks and leaves of field crops range from 0.04–2 per cent for nitrogen, around 0.1 per cent for phosphorus and 1.2 per cent for potassium.

The slow decomposition of the biomass in soil releases the nutrients to crops that are growing at the time. When organic matter undergoes degradation/synthesis processes in the soil, an end product is humic material with plenty of cation exchange sites to retain and make available nutrients such as calcium, potassium and ammonium, as well as secondary and trace nutrient elements. Microorganisms are largely responsible for decomposition that goes on in the soil. Organic matter serves as a carbon source, or food material, for many of these important heterotrophic organisms that live in the soil. In this way it supports the overall population of soil flora and fauna that is important in many aspects of nutrient cycling.

Composting the organic matter prior to its addition to the soil has beneficial features. When organic matter undergoes decomposition, either in a compost heap or in the soil itself, there is rapid growth in the population of microorganisms that are able to make use of the added food source. In order to multiply and grow, the microorganisms also require nitrogen and other nutrients that are present in the soil and/or in the added biomass. For this reason, if fresh biomass is added directly to soil, there can be a temporary decrease in nitrogen levels available for plant growth, a process called immobilisation. It is only after the decomposition is substantially complete that the microbial population begins to die off and the immobilised nitrogen is released back into the soil. Prior composting of the plant material can minimise the nitrogen immobilisation problem. If the organic matter has already been converted into a partially decomposed and somewhat stable form, little nitrogen will be taken up when it is mixed with the soil, and the soil immediately enjoys the full benefits of the added biomass. Successful composting, however, requires a good supply of air and water, and care needs to be taken that nutrients are not lost either by leaching or by volatilisation. In many cases then, it is better to leave the fresh organic matter on the soil or to plough it under.

One of the positive features of no-till agricultural practices is that some of the stalks and leaves as well as the below-ground components (roots) of plants are left undisturbed on and in the soil. There they assist in holding together the surface soil, so that it is less susceptible to erosion by either wind or water. The surface litter and severed root materials slowly degrade and disperse in the soil, providing the benefits noted above.

The organic matter of most soils accounts for a small percentage of the total soil mass. For many agricultural soils, organic matter content is greater in temperate regions of the world than in the tropics (particularly the arid tropics). This is due to the slower rates of degradation where the climate is cooler. Although some of the organic matter is degraded and resynthesised into fairly stable humic material, the ultimate degradation product is carbon dioxide.

The benefits of maintaining the organic matter level in soils are well known, but rapid oxidation to carbon dioxide is not desirable, both because the organic matter cannot serve its purpose of enhancing soil properties and because the carbon is released into the atmosphere without having served any useful function. Earlier, the benefits of improving soil properties by growing soil-building crops and incorporating their biomass into the surface material were suggested as the subject of indicators that can be used to assess aspects of stability.

## Biomass as fuel

A question of increasing interest is whether biomass should be grown for the purpose of producing fuels. There are sustainability implications in the environmental, social and economic spheres that bear on this question. The global proven reserves of the fossil fuels—coal, oil and natural gas—have an energy content of approximately 36,000 exajoules ($3.6 \times 10^{22}$ J). This is the same energy content as in 6 trillion barrels of oil. It is estimated that the net annual photosynthetic production of biomass is 7 billion tonnes ($7 \times 10^9$ t), and this has an energy content estimated to be 100 exajoules ($1 \times 10^{20}$ J). Global annual total energy use is 330 exajoules (EJ), about three times the net primary production of biomass.

About 90 per cent of the production of biomass is in the form of trees, predominantly in the world's forests, with most of the remaining 10 per cent being generated in croplands. It is clear that this organic matter is potentially an important resource that can serve a substantial portion (though far from all) the world's fuel needs. In fact, at present such biomass is an important source of energy in many countries, especially in low-income countries. In rural areas of these countries, agricultural wastes including straw from grains, cotton sticks and other secondary products of food production, along with wood from various trees and shrubby bushes, and animal (particularly cattle) manure are all widely used for domestic cooking and heating purposes.

There are attractive features of using biomass in this way. For one thing, it has the potential to close the soil carbon/biomass carbon/carbon dioxide cycle within a local area. As the biomass is photosynthetically

### Indicators for Assessing Agricultural Sustainability 177

**Figure 3.4**
A simplified view of the carbon cycle in an agricultural setting

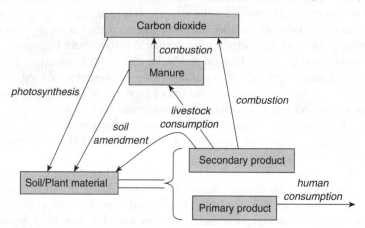

generated, carbon dioxide is removed from the atmosphere to form the mostly cellulosic and starch components of biomass. When the biomass is burned, the carbon combines with oxygen in the atmosphere and carbon dioxide is released back into the atmosphere (see Figure 3.4). Although the two steps in this cycle are not completely efficient, to a large extent this sequence of processes means that there is no net gain or loss of atmospheric carbon dioxide. This is in contrast to the burning of carbon-based fossil fuels that have accumulated their carbon content over many millennia in the past. When fossil fuels are burned, their carbon is newly added to the atmosphere with no simple immediate way to draw it back into a solid or liquid form. Another advantage of burning locally available biomass is that it is usually produced on the farm or in uncultivated areas that are in close proximity to the location where it is used. Without any commercial intervention, it then becomes possible for inhabitants of even remote and inaccessible areas to obtain a supply of fuel as needed.

But there are major disadvantages in the use of biomass as fuel. For one thing, we have shown that some forms of agricultural secondary material are excellent as fodder—some primarily as a source of carbohydrate and roughage, but others containing substantial concentrations of essential nutrients. The above ground portion of the groundnut plant was presented as an example of an excellent food source for animals, and it would be a very inappropriate use, both economically and in terms of nutrition, to have it burned instead of consumed. A second disadvantage is that combustion of biomass rules out its application as a soil amendment (except for the ash, which is in any case all too often discarded). Although on a weight basis organic matter is a minor constituent of most agricultural soils, it may still contribute to improvements in soil physical, chemical and biological quality to an extent that is much more significant than the small amount present would suggest. Finally, another problem with the use of biomass as a fuel is that it is far less convenient to use than liquid or gaseous fuels. It can also easily create major health hazards, especially for the homemaker/housewife, who is present near the combustion source much of the time when cooking is being done.

The inconvenience involved is also not trivial. In extreme cases, collection of biomass requires major effort to travel long distances when local sources are not available. This social (largely gender-related) problem has been well documented by many writers. When dried animal manure is used as fuel, it requires collection of the manure, mixing with water and other plant wastes, and forming 'cakes' by hand, after which they are dried—a time-consuming, unhealthy and unattractive series of operations. Biomass can also be difficult to ignite and difficult to control in a way that desired temperatures can be maintained for the required time. But perhaps most serious is the potential for creating respiratory problems when the combustion is carried out in an enclosed, improperly vented area. And this is very often the case as will be described below.

Biomass need not be burned directly, and there are a variety of technologies—some appropriate only on an industrial scale and some adaptable to small individual or community operations—by which biomass can be converted into more convenient and/or safer forms.

1. Pyrolysis (thermal decomposition in an inert atmosphere) of wood or other biomass to form charcoal is carried out by a variety of processes on a small or large industrial scale. The products formed from 100 kg of wood are typically:

    | | |
    |---|---|
    | Charcoal | 30–50 kg |
    | Pyrolysis oil | 30–50 kg |
    | Pyrolysis gas | 15–20 kg |

    Each of these products can have a number of useful functions. The charcoal is useful as a relatively clean-burning fuel with

Indicators for Assessing Agricultural Sustainability 179

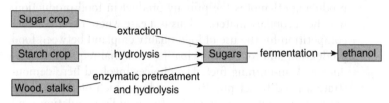

Figure 3.5
Production of ethanol from various feedstocks

a substantially higher heat content than the original wood, on a weight-for-weight basis. Pyrolysis oil has potential as a substitute for industrial fuel oil. The pyrolysis gas is usable at the production site itself as a low-grade heat source for the pyrolysis itself.

2. Gasification is an industrial process in which the biomass is converted into a gaseous form by partial oxidation at high temperature. In the multi-step process a mixture containing carbon monoxide, hydrogen, hydrocarbons like methane, and other inert gases is generated. This producer gas can be used in local heating or electrical generation applications. The process is quite inefficient and generally requires being carried out on a fairly large scale.

3. Fermentation of biomass materials in a stirred tank fermenter with the addition of an appropriate culture and maintenance of appropriate growth conditions is a standard procedure for producing ethanol. Easily fermentable substrates, particularly sugars and starches, are most appropriate, whereas the lingocellulose materials that make up a large proportion of the composition of straw and wood require enzymatic pre-treatment and hydrolysis in order to release the easily fermentable sugars (see Figure 3.5). There is active interest in developing appropriate cultures, enzymes and reaction conditions so as to maximise the efficient production of ethanol from these common waste materials. Separation of ethanol from the fermentation liquor is one of the difficult and expensive challenges in the overall process. Production must be carried out on an industrial scale. The ethanol produced by fermentation is a very clean burning fuel and one that is convenient to handle.

Because of the relative ease of ethanol production from sugar (cane and beets) and starch (maize) crops, there has been increasing interest in using these crops as feedstocks for biofuel production. In other words, the most desirable feedstock for producing ethanol is the primary product in food production, not the secondary material. This is a case where there is direct competition for the use of high quality cropland between food production and for the principal raw material required to produce a clean-burning fuel. The choices involved here demonstrate a need to set priorities regarding which of the possible products is more important for the world's population, and which of these choices meets the criteria of sustainability.

4. Biogas production on a large or small scale is one of the most promising options for use of the secondary material produced by agriculture. In contrast to most other technologies for conversion of agricultural wastes, biogas production can be carried out in large-scale industrial processes or on site at the level of a community, including, in some cases, at the level of an individual farmer. Essentially, biogas is a mixture of methane and carbon dioxide produced in a complex (but potentially easy to carry out) series of processes by the anaerobic (absence of air) fermentation of a variety of biomass materials. The quality of the biogas produced depends on the relative concentrations of methane (the fuel) and carbon dioxide (which is not combustible). The rate and efficiency of the methane production requires careful control of conditions, including the type of raw materials supplied to the bioreactor. Feedstock (biomass plus water) quality depends on having a sufficient supply of nutrients like nitrogen and is also limited by the content of lignin in the cellulosic material supplied to the system. The high lignin content of wood makes it a poor substrate. Straw is somewhat better but when used on its own requires pre-treatment to free the carbohydrate material from its association with lignin. When, however, the straw is used as bedding material for animals, the combined physical pre-treatment of trampling by the animals and biological pretreatment associated with its mixture with high-nutrient content manure converts the straw into an acceptable raw material. Green plant material is also a highly suitable substrate for direct introduction into the biogas plant.

The technology of biogas production can be sophisticated and complex (involving flow-through fluidised bed reactors, with careful environmental controls) or relatively simple. Simple batch systems are appropriate for use by individual or small groups of farmers. One limitation in designing small-scale systems is the need for some measure of temperature control. Optimum temperatures are about 35°C, and production of biogas falls off markedly when ambient temperatures fall below about 15°C, so that some kind of auxiliary heating system may be required in areas with cold climates.

Biogas produced in a small scale reactor can be piped into a home where it serves as a clean-burning fuel[11] to be used for heating and cooking. The problem of particulates being produced when biomass is directly burned is almost completely eliminated. The other product of the biogas production reaction is a slurry or sludge that remains in the reactor. After reaction, this residual sludge, representing about 30 per cent of the mass added to the reactor, can be pumped out and applied to the soil. It contains most of the nitrogen, phosphorus, potassium and other mineral nutrients of the original feedstock. The organic material that remains is made up of lignin-related material from the added plants and bacterial cell components that have grown during the fermentation processes. The sludge can be added to soil as a clean-smelling and substantially pathogen-free amendment that supplies nutrients and is a source of organic matter, enhancing the physical and chemical properties of the soil.

## Sustainable choices for use of secondary material from agriculture

It is clear that secondary material produced in the course of producing a food or other commercial crop is not a waste but should be considered to be a potentially valuable resource. To establish a sustainable agricultural system then, serious consideration of how to best use this resource is required. Considering the many possible situations, there will not be a single best choice that applies in all cases.

There are, however, some clearly undesirable and unsustainable options. Discarding the secondary material haphazardly or in a landfill is the most obviously unsustainable practice. Burning it on the field does provide some nutrients in the ash that remains, but the organic component is lost—and significantly converted into the greenhouse gas carbon dioxide—without providing any compensating benefits. Also, combustion of mixed biomass in an open field is usually a major source of harmful particulates that seriously degrade the surrounding air quality.[12]

Burning biomass inside the home for purposes such as cooking is a very common practice. It is usually a very inefficient process, with efficiencies as low as 10 per cent or less (as measured by the ratio of heat captured in the cooking vessel to heat value of the fuel) and also contributes to atmospheric pollution. In terms of human (particularly women's) health, when the cooking is done in an improperly vented or unvented enclosed area, it must be categorised as a highly unsustainable activity. Interestingly, there is a positive feature to the practice in that the smoke controls insects in the living area. Over the years and in many countries, newly designed stoves have been developed that burn more cleanly and provide more useful heat per weight of biomass than do traditional stoves. Where it is necessary to burn biomass directly then, it is incumbent that there be proper ventilation and that the most efficient stove be employed.

The ash, typically 1 to 10 per cent of the dry weight of the biomass, should also be considered to be a useful resource. Its high content of nutrients, such as potassium and silicon, makes it a useful soil amendment for application in fields, orchards or kitchen gardens. Having pointed out the unsustainable features of direct burning of biomass, it is important to reiterate that the desirable features include its ready local availability and the fact that combustion of locally produced biomass can be part of a substantially closed cycle of carbon dioxide sequestering and release.

Perhaps the most sustainable option, when it can be applied, involves a combination of some of the possible uses (see Figure 3.6). It goes without saying that a first priority for any appropriately nutritive biomass would be to use it as animal fodder. The animal waste, along with bedding, other less nutritive biomass, food scraps and human waste together then will provide an excellent feedstock for a biogas reactor. The methane is used for cooking and the sludge as a soil amendment.

The following quantitative comparison (based on typical data about composition) of options for use of biomass is instructive:

**Figure 3.6**
Sustainable strategies for use of agricultural 'wastes'

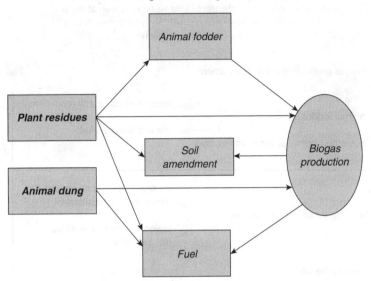

- One tonne (dry weight) of biomass, when added directly to the soil, supplies organic matter, plus approximately 6 kg of nitrogen, 1.5 kg of phosphorus and 3 kg of potassium.
- One tonne of biomass if directly burnt provides heat equivalent to 600 litres of kerosene.
- One tonne of biomass, after converting to biogas and sludge, produces gaseous fuel that could provide heat equivalent to about 400 litres of kerosene, and a soil amendment consisting of about 300 kg of organic matter, with nutrient content similar to that in the original material.

## Indicators related to the use of secondary material

As is evident from the discussion above, there is a range of options for the use of the secondary products of agriculture. Within that range, one could perhaps establish a hierarchy of values for the various applications. Such a hierarchy might take the form shown in Figure 3.7.

**Figure 3.7**
A hierarchy of possible uses of agricultural secondary products ranging from most sustainable uses at the top to least sustainable uses at the bottom. The dividing line separates generally acceptable from generally unacceptable applications

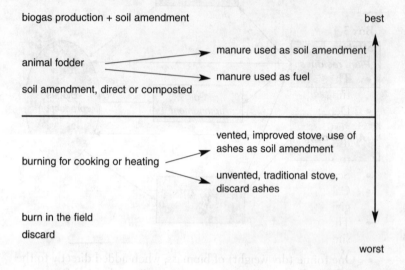

Within this sequence, adjustments are required to suit each situation. For example, cooking using a fuel-efficient stove in a properly ventilated setting, with application of the ash in a kitchen garden, could be acceptable when using low-nutrient content materials. Or, on the negative side, carelessly composted wastes could lead to unnecessary loss of nutrients and poor quality product for use as a soil additive, and would therefore be an unacceptable choice.

Taking into account all the possibilities, a broad response indicator related to efficiency of use of biomass would be to differentiate between the three lower categories in the hierarchy and the three upper categories. A response indicator would then be the percentage (scaled by dividing by 10) of total material used either as animal fodder, direct soil amendment or biogas production.

## Efs1 (Sustainable Use of Secondary Biomass):

Efs1s = (% sustainable use of secondary biomass)/10

> **Box 3.4**
> **Strategy for assessing efficiency**
>
> - The scope of the assessment should be defined (area and time to be covered).
> - Determine as many indicators as possible from those suggested for measuring water use efficiency (Efw1), energy use efficiency (Efe1 to Efe4), nutrient use efficiency (Efn1 to Efn3) and sustainable use of secondary biomass (Efs1).
> - Scale each of the chosen indicators to a common scale.
> - Calculate the average within the three sub-category values, giving a measure of each of these essential input values.
> - The three sub-indices can then be averaged to produce a single stability index.
> - Where appropriate, calculate additional indicators for efficiency.

## 3.4 Durability

*Durability*  Any crop production process is from time to time subject to stresses of various types—stresses such as those due to water or to pests. Sustainable systems are intrinsically resilient in the face of such stresses.

Durability, often referred to as resilience, is fundamentally defined as the ability of the agricultural system to resist stress and maintain a good level of productivity. A comparison of the meaning of durability and stability is instructive. Stability, as we have indicated, is a relatively long-term measure—maintenance of productivity over many years, in fact indefinitely. In contrast, durability considers maintenance of productivity over a single season. After a crop is planted, it is subject to whatever environmental conditions obtain, and these conditions are not always ideal. The ability of the crop to thrive despite less-than-ideal conditions is characteristic of a durable system. The ability of a farmer to live comfortably in spite of various short-term setbacks is similarly a characteristic of a durable economic system. Durability depends on the specific crop(s) being grown, but more than that it depends on the agricultural system in its totality.

In some instances there is a connection and a potential conflict between stability and durability. In order to ensure a stable food supply, humans have adopted a variety of usually energy- and resource-intensive farming practices. The artificial inputs, as long as they exist, maintain the agricultural system at a level that can be far from an equilibrium natural situation (see Figure 1.6). The result of this is that crop production remains stable as long as the inputs are fed into the system without adverse consequences. At the same time, however, the system may lose its intrinsic ability to resist stress. If it becomes impossible to continue the artificial support, the entire production system can suffer major declines, even collapse.

Irrigation is an example of this. When productive stability is maintained by having an assured water supply, crop management systems are developed that rely on the continued availability of the irrigation resource. However, should the water supply become degraded or unavailable, the unmodified methods of crop production may become seriously susceptible to environmental stresses. Similar loss of durability can result if agriculture becomes dependent on other resources such as ready availability of fossil fuels or other chemical inputs.

The same can be said for social and economic systems related to agriculture. If government support programs encourage the growth of a single crop by making its production highly lucrative in comparison to other options, there is a temptation to maximise these benefits, at the expense of diversification. Often such support is transitory, depending as it does on policy decisions made far from the local scene. With such support removed, farmers who specialised in one area suffer serious consequences in terms of income loss.[13] Farmers who have developed other options are however better able to withstand the imposed stress.

Because of these connections, it is important that indicators that fall within both the stability and durability categories be carefully

### Figure 3.8
The nature of the response of an agroecosystem to an imposed stress defines the durability of the system

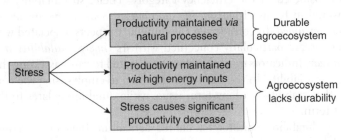

developed and assessed. If, in fact, the system is found to be very stable, but also intrinsically less durable, serious questions should be asked about the means of achieving stability. The overall conclusion in such situations may well be that there are important issues to be asked about the long-term sustainability of the system.

A durable cropping system, then, is one that remains productive during a given season in the face of one or more imposed stresses (see Figure 3.8), and which remains productive without the intervention of synthetic and high-energy inputs. In other words, it uses its own built-in natural resources, or at least those that do not impose a highly unnatural regime to resist the stress. We will consider the two most common types of stresses—water surplus or deficit and pest problems—in developing indicators that assess the durability of an agricultural management system. We will also look at economic impacts and the stress-response issues that farmers face in that sphere.

### Durability in the face of water stress

Water stress can be associated with excesses, sometimes leading to flooding; but it is more commonly linked to water deficits associated with drought conditions. It is the latter stress upon which we will focus. It may be recalled that water quantity issues played a part in developing stability indicators. Under the stability category, the most important parameter was total rainfall and its variability over time. In that context, the requirement for continued reliable rainfall was to ensure that *the resource would be indefinitely available* in the form of precipitation itself and as the source of water that keeps rivers flowing, lakes at a constant level and a stable water table. Water issues were also considered in the efficiency category. There, sustainability was measured in terms of *maximising crop output against the amount of water consumed*. In considering the durability aspects associated with water, we are particularly concerned with its *timely availability in a given year*. Indicators must therefore be directed towards assessing this type of variability. In this context it is worth re-emphasising the concept that stability is a long-term issue, while durability relates to the short term.

One durability indicator for rainfed agriculture therefore measures variability in rainfall during the critical months of crop production—for grain crops, typically at the early stages of their establishment and then at the flowering stage. This is done by calculating the coefficient of variation of rainfall during these critical months over a period of at least ten years.

### Dw1 (Critical Month Rainfall for Rainfed Areas):

Dw1 = Average value of the coefficient of variation of rainfall in the two critical periods (plant establishment and flowering) over a period of years.

Larger values are indicative of unsustainability. An alternative to this measurement is to determine the number of years (e.g. out of 10 years) when rainfall is sub-critical in either of the two critical periods.

### Dw2 (Critical Month Water Availability for Areas with Sources of Irrigation)

Dw2 = Average value of the coefficient of variation of water availability in two critical periods (plant establishment and flowering) over a period of years.

This indicator is closely related to Dw1. The variation in canal water availability, for example, ultimately derives from variations in rainfall. This yearly variation can be quite large, especially in the tail end of irrigation command areas, compromising the resilience of farming operations there. As in the case of Dw1, an alternative is to measure the number of years when there is inadequate water during the critical periods of the growing season.

One might assume that these two pressure indicators deal with subjects that are beyond the control of individual farmers. This is indeed true in most cases, although appropriate planning can be helpful in minimising water stress to some degree. The construction of water harvesting structures, lining of distributaries and channels that bring canal water to the fields, and the implementation of effective user groups to ensure equitable sharing of irrigation resources are some of the means by which stress due to limitations in amounts of available water can be alleviated. Therefore it can be useful to design a response indicator of the type suggested here that evaluates initiatives taken to relieve water stress.

**Dw3 (Water Collection and Distribution Initiatives):** This response indicator calculates the percentage of farmers involved in initiatives to maximise water harvesting, minimise water loss and/or share water equitably. Scaling to give Dw3s can appropriately be done by converting the percentage value to a scale of 0 to 10 (for example, 37 per cent participation gives an indicator value of 3.7).

$$\text{Dw3s} = (\text{Percentage farmer participation in water harvesting and loss prevention})/10$$

It may appear that this indicator should not apply where rainfall is assured, meaning that there is no need for these initiatives. However, water harvesting is also of benefit in stability terms as it reduces erosion and sustains the groundwater resource. Lining artificial water courses with impervious material improves stability by controlling seepage that is a cause of salinisation.

### Stress related to pests

A second major stress affecting crop production is associated with pests: disease, insects and weeds. Here we must consider the use of pesticides as a means of controlling such pests. The issue of pesticide use is complex. One of the important components of the green revolution management package has been the application of pesticides to control insect, acarid, fungal and weed problems, and this strategy has been a major enabling factor supporting the substantial increase in global food production. Unfortunately, pesticides are never totally selective, and predatory species are killed along with the pest species. In the absence of natural control by predators, pest populations can increase exponentially to devastating numbers with sudden and catastrophic loss of crops. Resistance to pesticides is itself a problem, one that encourages more and stronger dosages, or the development of new, sometimes more toxic, pesticides. At the same time, the widespread use (and particularly the overuse) of pesticides has in places created severe stresses on the general environment, including on human health.

Related to this are the current issues surrounding the commercial development of genetically modified organisms (GMOs). This twenty-first century strategy towards plant breeding promises many benefits including a lessened requirement for pesticide use. For example, Bt-cotton resists the devastating effects of the bollworm and it is claimed that growing these strains of cotton can substantially reduce pesticide use. It is worth noting that 'conventional practices' for cotton management may require upwards of twenty sprayings of various pesticides during a single growing season. This is a very complex environmental, economic and social issue, and whether or not developments in this area are truly sustainable is something that requires the closest scrutiny. At these early stages, we will only say that the utmost caution should be exercised in testing and adoption of GMO technology.

Pesticides are by definition toxic materials and it is this toxicity that is at the heart of the problem. It is also necessary that they persist sufficiently long within the environmental situation where they are applied. These situations may include being in contact with various potentially reactive soil and plant materials under conditions of intense sunlight, rainfall, and/or wind. While pesticides are reacting and interacting to destroy targeted pests, they may also come in contact with non-target organisms—other plant and animal species. There is strong evidence that many of these synthetic organic chemicals can have toxic, mutagenic and carcinogenic consequences at every stage of their existence, from the manufacturing process through their application in various environmental settings to the point at which the agricultural products are used or consumed far from the site of application.

Using standard technology, manufacturing of pesticides is increasingly being carried out in low-income countries where environmental standards are relatively lax. The most infamous example of this is the Union Carbide plant in Bhopal, India, that was the site of manufacture of methyl isocyanate, a starting material for the manufacture of the widely used carbamate pesticide, carbaryl. In December 1984 there was a massive release of this chemical, resulting in the death of at least 3,000 persons and adversely affecting the health of at least 300,000 others. A number of factors were responsible for this disaster, but most of them can be summarised by saying that the company took advantage of a system where there were lenient safety standards and very limited control of manufacturing procedures.

Compounding the problems of increased concentration of manufacturing in low-income countries is the fact that large accumulations of pesticides banned by national and international treaties are still stockpiled at various locations in these countries. Many of the chemicals are highly persistent organochlorine compounds, while others are less persistent but very toxic members of the organophosphorous pesticide class. The environmental and health consequences of their long-term storage, often under less than ideal conditions, have never been assessed.

Application in the field is another area in the life cycle of pesticides where there can be serious impacts on the natural environment as well as on human health. There are many reported cases of non-target organisms being killed. For instance, beneficial species of spiders and beetles in rice, and coccinellids, chrysoperla and spiders in cotton ecosystems come under the influence of pesticide stress. There are also well-documented instances of the development of genetically transmitted resistance within the insect or plant populations being targeted by the pesticide. Resistance cycles as short as five years have been reported in Africa. The increasing resistance generates a need to create new, often more toxic chemicals, but in the interim period it is not uncommon that farmers feel obliged to resort to application of higher and more frequent doses of pesticides in an attempt to ensure the survival of a vulnerable crop.

Furthermore, the use of pesticides presents a health risk for the person carrying out the application. This is especially true where workers are not provided the opportunity for proper training in safe handling procedures, or with the equipment needed for safe handling. The World Health Organisation estimates that three million people each year suffer from severe symptoms of pesticide poisoning and one tenth of this number die as a consequence.

In terms of pest control, a durable system is one in which pests pose only minimal problems, either because serious crop pests are not found in that area or because pest management is effected using natural means. As we stated at the outset, durability is then dependent on the nature of the crop itself and also on the total management system employed on the farm.

When chemical intervention becomes necessary, it should be carried out in as environmentally cautious and benign a manner as possible. The chemical interventions are then only a minor part of an integrated pest management protocol. In this light, indicators of durability are designed as a measure of the level of chemical control required.

An indicator that measures the chemical response to pest stress is suggested here.

**Dp1 (Chemical Response to Pest Stress):** For a single crop, the indicator for chemical response to pest stress is calculated from the number of sprayings during the growing season. When an area that is under a common management system (an agroecosystem) is being studied, a response to pest stress indicator can be calculated after determining the average number of applications of pesticides per year for all crops; the average is then weighted by area.

$Dp1$ = chemical response to pest stress, measured by average number of sprayings in a specific agroecosystem, N

$$N = (\Sigma C_{ni} A_i)/A_t$$

(where N is the average number of sprayings in a specific agroecosystem, $C_{ni}$ is the number of times each crop (i) grown on the area ($A_i$) is sprayed. $A_t$ is the total area covered by the common agroecosystem.)

To illustrate this, consider an area covering 10 ha, where on 4 ha two crops are grown during two seasons in a single year. The first crop requires six sprayings and the second requires three sprayings. On the other 6 ha, a single crop is grown in that year and the crop is sprayed 7 times. The average number of sprayings on the total 10 ha area is therefore:

$$(4 \times 6 + 4 \times 3 + 6 \times 7)/10 = 7.8$$

To convert this number into an indicator, goalposts are selected, e.g., the optimum number (B) of sprayings as 0, and the poorest value (P) as 12.

$$Dp1s = \frac{P - N}{B - P} \times 10$$

Table 3.13
Sustainable methods of pest and disease control

| Sustainable methods of pest and disease control |
| --- |
| • Crop rotations and intercropping (of different species and genuses)<br>• Companion planting<br>• Use of resistant/tolerant varieties<br>• Use of alleopathic/antagonistic plants<br>• Use of physical barriers such as tree breaks or insect traps<br>• Use of biological controls such as predators<br>• Control of carriers<br>• Use of natural pesticides<br>• Hand picking |

*Source*: Modified from information obtained from Parrott and Marsden (2002).

The indicator value would then be calculated to be:

$$Dp1s = (12 - 7.8)/12 \times 10 = 3.5$$

Clearly, as in other cases, the value of the indicator depends on the goalposts selected. Nevertheless, in an area occupied exclusively by organic farms, the chemical response to pest stress indicator would always have a value of 0.

The indicator just described measures a chemical response to pest stress. This 'negative information' was scaled in such a way that a large number indicated fewer sprayings and a more durable system.

There are other responses to pest stress that do not make use of chemical inputs. Table 3.13 lists some of these. As in the case of water stress, these positive responses can be used for the creation of an alternate durability response indicator.

**Dp2 (Non-Chemical Response to Pest Stress):** This response indicator calculates the percentage of farmers involved in initiatives to control pests by non-chemical means. Scaling to give Dp2s can appropriately be done by converting the percentage value to a scale of 0 to 10 (e.g. 24 per cent participation gives an indicator value of 2.4).

$$Dp2s = \text{(Percentage of farmers using non-chemical pest control)}/10$$

Note that this indicator may not appear to be relevant where there are no perceived pest problems. Nevertheless, such initiatives may serve to prevent pest problems that might otherwise occur.

## Economic stress

A separate but related aspect of the durability equation is the economic issue. Crop failures associated with environmental stress lead to economic hardship. Where there are frequently occurring failures, the farming operation itself becomes unsustainable. This problem can be measured using an appropriate indicator.

**De1 (Years of Economic Hardship):** A simple state indicator can be developed by determining the number of years (N) out of 10 when net farm income falls below a certain level. Clearly the level chosen is different for each situation. It should never be less than the poverty level, but should ideally be substantially greater, supporting a comfortable standard of living. The value of the indicator is then given by

$$De1s = 10 - N$$

As a means of coping with environmental or economic stress, an alternative approach widely used throughout the world is for farmers to develop a diversified portfolio of activities. These may involve growing a variety of crops, particularly ones that are affected to varying degrees by the possible stresses that occur in that agroecoregion. Other activities beyond crop production add further to the diversity and therefore the resilience of the system. In addition to growing a variety of crops, these operations could include a whole range of separate activities, including animal husbandry, i.e., raising livestock for dairy or meat, bee-keeping and production of honey, forest-related activities, machine repair, etc. These activities not only add an additional source of income, but also provide an outlet for some secondary agricultural products, which we have seen as contributing to the overall efficiency of the system.

It is important in evaluating the significance of this indicator to be aware that there are at least three, sometimes contrary, pressures, which are contributing factors to the population involved in rural non-agricultural employment. One obviously negative factor is distress-driven movement of persons into often low-paid and temporary activities, due to loss of more stable employment in the agricultural sector. The more positive reasons arise either as a direct result of agricultural growth, with attendant spin-off opportunities connected with improved agriculture, or because of government-encouraged employment generation initiatives or provision of services in rural areas.

Indicators that evaluate response to actual or potential economic pressures can be developed. One strategy to minimise the possibility of economic problems due to crop failure centres around development of diverse activities in the total farming operation.

**De2 (Agricultural Diversification):** This response indicator determines the percentage of farms where diverse activities, aside from crop production, contribute a significant fraction (say at least 30 per cent) to the average yearly income. The indicator can be directly scaled (De2s) from 0 to 10 (percentage/10).

$$De2s = (\text{Percentage of farm families involved in diversified activities})/10$$

In dealing with issues of durability, both in terms of trying to assess the resilience of a farm or agroecosystem, and in terms of strategies for responding to stresses, the opinions and advice of older persons can be invaluable. Veterans who have practiced farming in a particular area over many years have observed the whole range of stresses. Through their keen powers of observation, they can often predict problems before they arise. Besides, during the period before the wide availability of the many inputs that have become so accessible, they were often able to develop relatively benign strategies to counteract the stress.

---

**Box 3.5**
**Strategy for assessing durability**

- The scope of the assessment should be determined (area and time to be covered).
- Determine as many indicators as possible from those suggested for measuring water stress durability (Dw1 to Dw2),

---

*(Continued)*

## (Continued)

> - pest stress durability (Dp1 to Dp3) and economic durability (De1 and De2).
> - Scale each of the chosen indicators to a common scale.
> - Calculate the average within the three sub-category values, giving a measure of ability to withstand water, pest and economic stresses.
> - The three sub-indices can then be averaged to produce a single durability index.
> - Consider the implications of the individual values of the indicators, and take into consideration advice from experienced persons.
> - Note the connections between durability and stability and ensure that the measures to enhance stability are not compromising the resilience of the system.

## 3.5 Compatibility

*Compatibility*   Sustainable agriculture should fit in with the human, social and natural environments where it is located, maintaining and enhancing the health of these environments.

In a broad sense, compatibility refers to the ability of an agricultural system to fit in with the biogeophysical, human and socio-cultural surroundings in which the system is placed. As was discussed in chapter 2, over the long period of the Earth's history, natural communities of living plants, animals and microorganisms have developed and existed outside the control and influence of human beings. The nature of these systems depended on the fundamental geochemistry and climatic features of the area, but each system was also influenced by all sorts of interrelated life processes. Over time, a particular area developed common and definable features and this could be called an ecosystem. Some sort of equilibrium or steady state was achieved, with production and consumption in approximate balance.

But one should not think of such ecosystems as being static and unchanging. Long before the advent of humans, from time to time periodic local and global events—fires, volcanoes, earthquakes, infestations, changes in climate—disturbed every ecosystem of the world, and there followed periods of adjustment to the new environment, until a new temporary equilibrium situation was achieved. With the evolution of the human species, opportunities for altering the environment became ever more frequent and significant. Now there is virtually no place on Earth that is not being influenced in a major and changing way every year. It is therefore important to think of humans as part of each ecosystem in which they reside.

Humans live in communities and, like natural biological ecosystems, the human social systems that have developed over large areas and in small pockets around the world each have unique and special characteristics. To some degree, the features of every social system are determined by the fundamentals of location, climate, resources, etc. Added to these are influences of religion and history. Like natural ecosystems, human social systems can be stable, in equilibrium, over periods of time. But they too undergo change, slowly as ideas and practices evolve from within, or suddenly, often due to catastrophic events of nature or events caused by humans themselves. One thing is certain—humans and nature are always intertwined and each influences the other in ways that are sometimes, but not always, recognised. It is these connections that should be reinforced in a positive way, and it is also these connections that come under the definition of compatibility. Clearly, this definition opens the possibility of considering a vast array of issues and could lead to demanding criteria for sustainability. In an attempt to measure the degree to which agricultural practices are compatible with humans and human society, and the degree to which human society is compatible with agriculture, we will limit ourselves to three aspects of compatibility.

### Human (as individuals) compatibility

When we talk about being compatible with human beings, we are referring primarily to agricultural practices that support their physical well-being. Clearly agriculture, involving as it does knowledge and planning, human labour, and work with living plants and animals is an intrinsically healthy and fulfilling activity. Nevertheless, the degree to

### Table 3.14

A selection of safe drinking water parameters as defined by the World Health Organisation (WHO 2003). These are guidelines that, if exceeded, could cause serious adverse effects. In addition, smaller concentrations ingested over an extended period of time may create chronic problems. Detailed guidelines are available for many additional parameters—other metals, non-metals and a large range of specific organic compounds, including pesticides

| Drinking water standards (World Health Organisation) | |
| --- | --- |
| Arsenic | 0.01 mg/L (milligrams per litre) |
| Boron | 0.5 mg/L |
| Chlorine | 5 mg/L |
| Fluoride | 1.5 mg/L |
| Nitrate | 50 mg/L |
| Nitrite | 3 mg/L, 0.2 mg/L for long term exposure |
| E. coli or thermotolerant coliform bacteria | Must not be detectable in any 100 mL sample |
| Total coliform bacteria | Must not be present in 95% of samples taken throughout any 12-month period |

More detailed guidelines are available for specific microorganisms.

which true fulfilment is realised in the lives of individuals and societies varies from situation to situation. For many persons the extent and nature of physical labour are the major factors that define whether or not agricultural practices are fulfilling. One should not romanticise the benefits of such activities, as anyone who has experienced the backbreaking work required for transplanting rice seedlings while ankle-deep in the wet clay soil of paddy field under unrelenting sunshine can testify. Technology and science that relieves some of the hard physical labour is a good thing, but it should be kept within a perspective that also considers other issues of sustainability. The issue of employment, for example, is important. With more than 70 per cent of the population of a number of countries directly involved with agriculture, their sudden displacement by machines could seriously disrupt the social cohesiveness of the country.

There are also measurable health issues to consider under the heading of compatibility. Water quality is one such issue, one that was also a concern within the stability category. In terms of compatibility, however, the important feature is to ensure that agricultural practices do not compromise the quality of local water that is to be used for domestic purposes, particularly for human consumption. The requirements for safe drinking water are considerably different from the requirements for good quality irrigation water. Table 3.14 tabulates a small number of important parameters within the Drinking Water Guidelines defined by the World Health Organisation (WHO 2003).[14]

These and any other specific contaminants for which measurements can be made on domestic water supplies should be included in assessments of the sustainability of agricultural practices. Some of the potential contaminants, such as fluoride, are almost always derived from the parent bedrock/soil, and the extent to which they are present is only slightly or not at all influenced by agriculture. Others, like nitrate (derived to a significant extent from run-off and leachates associated with nitrogen-containing fertilisers) and *E. coli* (often derived from livestock manure) are clearly agriculture-related. Pesticides and their degradation products that find their way into drinking water supplies are a class of compounds that must be of serious concern in rural settings. Unfortunately, analysis of the very small quantities that may be present (yet still potentially toxic even at low levels) is difficult and is carried out only in very well equipped laboratories. A third category of contaminants includes arsenic and other elements/compounds that are naturally present to varying degrees in bedrock and soils, but whose concentration in drinking water can be influenced by agricultural practices. This is a particularly serious issue in several places around the world, most notably in Bangladesh and Eastern India, and it is the subject of ongoing research.

In the present situation, we will look at indicators that focus on the second category, i.e. indicators that are connected with contaminants having a well-defined agriculture-associated origin. Pressure indicators based on nitrate and *E. coli* are examples.

**Ch1 (Nitrate Concentration in the Drinking Water Supply):** The nitrate concentration ($C_N$) in the drinking water supply is measured in milligrams per litre (mg/L), which is numerically equivalent to ppm (parts per million). The WHO upper limit guideline for nitrate is 50 mg/L. The optimum value can be taken as a concentration that is not detectable, or 0 mg/L, meaning none is detected. To scale the measured value, use:

$$Ch1s = (50 - C_N)/5$$

In this way, a sample of water containing 10 mg/L nitrate gives an indicator value of $(50 - 10)/5 = 8$.

**Ch2 (*E. coli* in Drinking Water):** The percentage ($P_E$) of samples in which *E. coli* is detected is determined. Since even a small number of samples is problematic, the goalposts should be compressed to a small range. For example, one might choose the optimum value as 0 per cent, and the poorest value in the measurement as 10 per cent of the samples being found to contain *E. coli*.

$$Ch2s = 10 - P_E$$

For a series of samples of which 93 per cent (percentage of contaminated samples $P_E = 7$) are free of the bacteria, the indicator is then calculated as $(10 - P_E) = (10 - 7) = 3$.

Similar pressure indicators related to health aspects of compatibility can be evaluated for any parameter that is associated with agriculture and is a potential problem in a particular situation. Note too that while salinity of water was treated earlier as a problem related to irrigation, it is also a health and aesthetic issue in terms of water used for drinking.

State indicators can also be developed to measure compatibility. Gastrointestinal problems are themselves a frequent indication of poor quality water. An indicator based on frequency of these problems is as follows:

**Ch3 (Water Related Illness):** The percentage of people (P) who have experienced gastrointestinal problems requiring medical treatment in the past year is evaluated. Setting best and poorest goalpost values at 0 per cent and 20 per cent respectively, the indicator is scaled to give:

$$Ch3s = (20 - P)/2$$

If 2.5 per cent of the population has suffered from these problems, the indicator value is $(20 - 2.5)/10 = 8.8$. This indicator has the limitation that gastrointestinal problems can arise from other sources as well, such as contaminated food or generally poor hygiene. It could therefore be an overestimate of the specific problem.

**Ch4 (Biocide Indicator):** As a predictor of possible problems associated with pesticide use, a rather complex biocide (pesticide) indicator has been developed that takes into account the amount of pesticides used in a given area, the pesticides' toxicities and their persistence in the environment. High values of this biocide indicator point to the possibility of adverse human health effects for farmers or their families who live near the fields where the pesticides have been applied.

The biocide indicator (Jansen et al. 1993) is defined as follows:

$$Ch4 = 1/Y \, \Sigma\Sigma A_{a,b} \, AI_b \, TOX_b \, DUR_b$$

where Y is the length of cropping period ($1/Y = 1$ for sugar cane and banana, for cotton = 2, for other crops = 3), a is the number of applications of each pesticide b, A is the amount of each pesticide added (mg), AI is the fraction of active ingredient, TOX is the toxicity (mg/kg$^{-1}$) of each pesticide measured as $LD_{50}$ (lethal dose, 50 per cent—the amount of toxic material, in this case provided orally to experimental rats and measured as mg of pesticide per kg body weight of rat, that has been shown to kill 50 per cent of the population), DUR is the half life, $t_{1/2}$, in days. These data are available in tables of environmental properties of chemicals. The index should be combined with information about how much of each pesticide is used, to give a weighted average for individual crops in particular areas.

### Ch5 (Groundwater Ubiquity Score) (GUS):

Similar to the biocide indicator is the groundwater ubiquity score (GUS) (Gustafson 1989). The GUS uses standard information about persistence and mobility ($K_{OC}$) of pesticides to generate a value that can be used to predict the possibility of the pesticide migrating into groundwater. This indicator provides important information in agricultural areas where groundwater is used for domestic purposes. The GUS does not take into account either the amount of pesticide or its toxicity.

$$Ch5 = GUS = \log_{10}(t_{1/2}^{soil}) \times (4 - \log_{10}(K_{OC}))$$

where $t_{1/2}$ is the half life (days) of the pesticide in soil/water systems and $K_{OC}$ is the organic carbon/water partition coefficient, both available from tables of environmental properties of chemicals.

GUS values greater than 2.8 indicate pesticides that are highly mobile and therefore have the potential to leach into groundwater, while values less than 1.8 are obtained for pesticides that will persist in the surface of the soil.

The above two indicators may be useful specifically for environmental research purposes, but they obviously require information from specialised technical sources for their proper evaluation.

**Ch6 (Protected Water Supply):** Response indictors related to human health take into account protective measures that have been utilised to counteract problems that could result from agricultural practices. For example:

Ch6s = (percentage of people with access to protected water)/10

While valuable in some situations, the significance of the indicator depends on the meaning of 'protected'. In many cases, measures taken to protect the water supply may be inadequate, especially in terms of chemicals that are readily leachable. Some means of advanced water treatment, often quite elaborate, may in fact be required and will rarely be in place in rural situations.

## Compatibility with cultural practices

This is a particularly complex issue and difficult to evaluate. Around the world there is a variety of ways of looking upon the position and role of human beings and their way of life within the environment. Some societies are particularly sensitive to the needs of all living things, considering humans as but one of many species, and one that must respect as equals other life (specifically animal) forms. It is easy to see that this view is compatible with much of current environmental thinking, but it is not without problems. When societies put specific human needs, e.g., needs for nutrition or hygiene, as secondary to certain practices that arise from their deep respect for all life forms, the quality of human life sometimes sinks to levels that should be considered to be unacceptable.

At the other end of the spectrum, there are societies whose broad cultural underpinning is largely human-centred—where humans have assigned to themselves the right to make decisions about all other life forms. In many cases this approach to life ends up promoting an exploitative, consumption-oriented lifestyle, wherein surroundings are only respected when they serve the needs of humans. However this view is not without its merits. Because it elevates humans to a special status, it can be the basis for demanding that humans take special responsibility for all that makes up the world. Stimulated by this mandate, scientists have developed many of the concepts of ecology; where agricultural practitioners have put these ideas into practice, great strides are made towards true sustainability.

To say that sustainable agriculture should be compatible with cultural practices then has to take into account each particular setting in which farming is carried out. In the type of society first described, this means treating agriculture in all its aspects as yet another activity of nature, respecting the seasons, working within constraints and accepting benefits that each environment provides. This philosophy should also provide the impetus to enhance life in its various forms through promoting growth and productivity, without indulging in practices that harm other life forms within the agroecosystem.

Where the second philosophy prevails, agriculture can be thought of as another opportunity for humans to fulfil their destiny as providers of good things. This too can mean treating the whole Earth as a garden that needs careful tending, recognising the qualities of every individual component and the supportive or destructive relations between each one. Care should then be taken that delicate balances are not disturbed by intrusive human activities.

Perhaps these subtle yet essential issues cannot be encapsulated within a collection of numerical indicators. It, however, behoves those who are working toward holistic evaluations of sustainability to encourage all those involved in agriculture to develop a supportive view of agriculture within their society.

## Compatibility with the biogeophysical surroundings (Biodiversity)

Biological diversity and complexity are important components of a sustainable agricultural system. A diverse agroecosystem provides environmental services which support sustainable agriculture itself but also provide benefits that go beyond the agricultural enterprise of the individual farmer. A biodiverse agroecosystem, one that incorporates trees and other large plants and provides connected areas for native species in their natural setting, is the source of many benefits:

- Wild species with special characteristics may be providers of medicines and other specialised products.
- The variety of species enlarges the range of habitats used by plant, animal and bird species.
- Permanent crop cover, especially trees, enhances both the quantity and quality of the water supply.
- The roots of crops hold the soil in place, minimising its movement by wind and water; this is especially important in sloping lands where erosion is a major issue.
- Maintaining biodiversity of crop species and their wild relatives is a means by which the store of germplasm for plant breeding endeavours is enhanced.

To achieve diversity and complexity, it is necessary that the farm and village be host to as wide a variety of plant and even animal species as possible. The augmented variety of species promotes diverse habitats and encourages increased complexity in population structure and dynamics within the agroecosystem. This in turn promotes provision of habitats that support pollinators as well as natural predators that operate against crop pests. In a diversified environment there is a reduced need for chemically based interventions and this results in the preservation of beneficial soil organisms that are involved in organic matter decomposition, nutrient cycling and promotion of good soil physical properties.

Agriculture that includes a variety of cropping also reduces the risk of catastrophic single crop failure associated with stresses like drought or pests, and is therefore a hedge against financial ruin (see Section on Durability Indicators). Intercropping is particularly effective in this regard since, if one species is adversely affected, the more tolerant plants can expand into the territory left vacant. Clearly, this is not possible where diverse crops are grown in separate fields. Taken together, it can be seen that promotion of diversity both in the overall landscape and in the cropped area contributes towards increased resilience of the agroecosystem.

There are further benefits however, as it is increasingly being shown that it is possible to gain significant yield increases in diverse cropping systems as compared to monocultures. This may result from a variety of mechanisms operating individually and together, especially related to more efficient use of sunlight, water and nutrients.

There are situations where it is well established that biodiversity contributes to increased productivity. Studies have shown that in the long term shade-grown coffee plantations can be more productive than their full-sun counterparts. The full-sun plantations may achieve very high levels of productivity for short periods of time but this requires high inputs of fertilisers and pesticides. Generally however, productivity declines after a relatively short period and the trees must be replaced. The result is an ongoing requirement for a cycle of high-cost and resource-intensive operations. In contrast, for example, coffee grown under shade remains productive over longer periods of time. Support for productivity comes from the interspersed trees, often leguminous varieties that supply a portion of the nitrogen requirements. In addition, the shade controls growth of weeds, reducing the need for herbicide inputs, and can also contribute to prevention of disease.

All of these issues, even though they are related to some of the earlier categories of sustainability, are best encapsulated by biodiversity indicators under the compatibility umbrella. Indicators of diversity include measures of crop species and varieties over space and time, use of management practices such as strip cropping, relay cropping, intercropping and employment of trap crops, and also measures of diversity, connectedness and quality of the surrounding natural landscape.

### Diversity indicators

There are well-established methods and indicators that can be used to assign quantitative values for measuring different aspects of diversity within an agroecosystem. The indicators may even, at one extreme, be used to calculate the overall diversity of a given area, including that associated with the surrounding natural vegetation. This would however require a detailed census of species within the uncultivated part of the landscape, a major undertaking in itself. For one thing, what should be included in counting species? Even if one restricts the evaluation to plants, there will in many cases be a great variety of plants, shrubs, and smaller plants, sometimes occurring in large numbers, sometimes as individuals. A comprehensive census requires extensive work by one or more trained botanists/ecologists.

Determining plant diversity within the cultivated portion of a landscape is, on the other hand, relatively simple. Nevertheless, this basic information is very useful as it allows one to describe diversity in a quantitative way within a whole range of agroecosystems. There will be diversity situations, ranging from broad areas under monoculture to areas where a large variety of crops are grown within a limited space. One important consideration in making diversity measurements is the issue of scale. Clearly, if one is counting species to measure diversity, the numbers will generally increase as one increases the area of land being considered. It is important therefore to define the areas under consideration, and these should be neither too large nor too small. Each situation is different, but for many purposes an area around 10,000 ha provides an appropriate scale within which to operate.

Of the many formulae recommended for assessing species distribution in different ways, some that can be used with crop data obtained either in the field or from remote sensing studies are as follows:

Richness — $N$
Proportional abundance
  (Shannon diversity indicator) — $S = -\Sigma A_i \ln A_i$
Evenness (Shannon evenness indicator) — $E = S/\ln N$
Relative dominance
  (Berger-Parker diversity indicator) — $D = A_{max}/A_{total}$
Similarity indicator (Jaccard's indicator) — $J = N_c/(N_a + N_b - N_c)$

where $N$ = number of different crops grown; $N_a$ = number of species in an area 'a'; $N_b$ = number of species in an area 'b'; $N_c$ = number of species common to areas 'a' and 'b'; $A_i$ = fractional area occupied by an individual crop; $A_{max}$ = fractional area occupied by the most abundant crop; $A_{total}$ = total of the fractional area occupied by all crops.

We will illustrate calculations involving these formulae using further data from an area in Karnataka State in South India.[15] The Tungabhadra Project (TBP) developed an irrigation command that provides an assured water supply over a region comprising 475,000 ha. In the head end of this region, there is extensive cultivation of rice (paddy), with a monsoon (*kharif*) and winter (*rabi*) crop grown on much of the land year after year. In the tail end however, where the amount of available water is considerably less, a greater variety of crops requiring only light irrigation are grown. Mapping the two regions using satellite-based remote sensing provided the following

### Table 3.15
Crop distribution in the head and tail ends of the Tungabhadra Project (Karnataka State, India) command area, in the 1998 and 1999 cropping seasons

|  | Head end | | Tail end | |
|---|---|---|---|---|
|  | % total area | % cultivated area | % total area | % cultivated area |
| Built | 11.5 |  | 2.6 |  |
| Water | 1.7 |  | 0.5 |  |
| Trees | 1.3 |  | 1.5 |  |
| Scrub | 4.6 |  | 14.4 |  |
| Rocky hills | 10.2 |  | 6.3 |  |
| Rice | 54.1 | 76.4 | 15.1 | 20.1 |
| Banana | 0.9 | 1.2 | 0 | 0 |
| Sugarcane | 0.3 | 0.4 | 0 | 0 |
| Sorghum | 7.1 | 10.0 | 23.3 | 31.5 |
| Cotton | 1.6 | 2.2 | 12.6 | 16.8 |
| Sunflower | 0 | 0 | 0.8 | 1.2 |
| Groundnuts | 0 | 0 | 0.6 | 0.8 |
| Fallow | 6.3 | 9.0 | 22.3 | 29.8 |

Source: From Thompson et al. (2001).

fractional information (expressed in percentage terms) in 1998–99 (Table 3.15).

Using this information, the various parameters are calculated in the following way:

*Number of crops* (N)
    Head end      $N = 5$
    Tail end      $N = 5$

*Proportional abundance* ($S = -\Sigma A_i \ln A_i$)
    Head end      $S = 0.79$
    Tail end      $S = 1.44$

*Evenness* ($E = S/\ln N$)
    Head end      $E = 0.49$
    Tail end      $E = 0.89$

*Relative dominance (calculated using area under cultivation)* ($D = A_{max}/A_{total}$)
    Head end      $D = 76$
    Tail end      $D = 32$

*Similarity index* ($J = N_c/(N_a + N_b - N_c)$)

Index used to compare common features of areas a and b   $J = 0.43$

Data from some of these formulae can be scaled into indicators that allow comparisons to be made. Of these standard parameters, perhaps the most useful is the proportional abundance parameter, which we can make use of to construct an indicator of crop diversity, a state indicator.

**Cb1 (Crop Diversity Indicator):** This indicator is based on the measure of proportional abundance:

$$S = - \Sigma A_i \ln A_i$$

This indicator has mathematically derived minimum and maximum values of 0 and infinity respectively. Zero corresponds to no diversity, i.e. a monoculture situation and, for example, 2.3 to a cropped area evenly divided among 10 species. On this basis, a diverse, but realistic (within an area of 10,000 ha) cropping system might yield a value of $S = 2$, and so this could be taken as the optimum value, while 0 would be considered the poorest value. Scaling then determines that the crop diversity indicator can be given by:

$$Cb1s = 10 \times S/2$$

For the examples cited above, the values of Cb1s are 4.0 for the head end situation and 7.2 in the tail end. These values reflect the differences in diversity in the upper and lower reaches of the command area. The importance of scale should be restated here. Clearly, for a single field where one crop is calculated, the index value will always be 0, whether the field is 1 ha or 100 ha in size. The value of the indicators calculated in this example were based on regions, each including several villages where a common set of agricultural practices were followed.

Two additional parameters that provide information about land cover diversity are also readily calculated from remote sensing or field-based data. These can also be used for development of other state indicators.

**Cb2 (Natural Environment Fraction):** This is the fraction of total area that is left in a natural state ($A_{natural}$); the natural state can be defined to include forest, meadow, uncultivated land of various other types, wasteland and water. The natural environment fraction, NEF, is given by $A_{natural}/A_{total}$. Using the present data:

> Head end      NEF = 0.14
> Tail end      NEF = 0.22

Goalposts are once again arbitrary; choices might be poorest value = 0 and best value = 0.4. The scaled indicator is calculated as:

$$Cb2s = 10 \times NEF/0.4$$

For the head and tail end situations here, the indicator values are 3.5 and 5.5 respectively. As with the crop diversity index, Cb1, this indicator also depends in the same way on the size of area over which the measurement is taken.

A crop rotation index is calculated by determining the average fraction of land in a given area that is planted to a different crop in subsequent seasons or years. For this calculation it is desirable to have several year's data; averages of pairs of successive values are then determined for all the fields in a specific area. The results reported below cover four successive cropping seasons, giving three individual pairings of data (not shown).

**Cb3 (Crop Rotation Fraction):** Crop rotation fraction = CRF = ratio of consecutive seasons when a different crop was planted on the same field over total number of consecutive seasons. Data is averaged for all fields within a given area.

> Head end      CRF = 0.32
> Tail end      CRF = 0.80

To compute the indicator, logical goalposts correspond to poorest values of 0 (no rotations) and 1 (rotations on all fields). Scaling then gives

$$Cb3s = CRF \times 10$$

This gives values of 3.2 and 8.0 for the head and tail ends respectively.

#### Figure 3.9
The figure on the left shows areas of remaining natural vegetation established with no connectedness, while the figure on the right shows the same area of natural vegetation where the existing natural remnants are located in a way that establishes ecological corridors. Shaded areas correspond to various types of native uncultivated vegetation

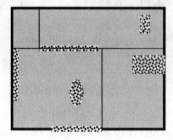 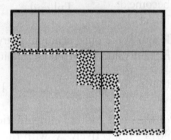

## Other diversity indicators

There are many other features of diversity that are worth measuring if one is undertaking a detailed study of this subject. Besides assessing species diversity within the surrounding natural environment as noted earlier, analysis of functional diversity contributes to a more complete picture. Functional diversity is a measure of the proportion of areas occupied by plant types that serve differing ecological functions. Types of groups to be considered might include trees, shrubs, small perennial plants, pasture land and annual crops. In a way this broadens understanding beyond the two-dimensional space-based views described earlier. By including functions of various species that share a given three-dimensional space, issues of the vertical element and lifespan of species also come into play.

Connectedness of the wild environment is another diversity issue. Where natural vegetation within a landscape forms a continuous series, it can support the maintenance of corridors and natural habitats for many animal and plant species. Connectedness in an agricultural landscape can be measured by determining the percentage of remnant natural vegetation that is connected within corridors (see Figure 3.9).

Besides the observable landscape parameters, diversity of species other than plants also provides useful information. This is true especially in the case of insects, since the numbers and varieties can be very large, contributing to the statistical validity of the measurement.

Also, depending on the situation, species can be either agricultural pests or predators of pests. In general, a diverse population provides some measure of control over the possibility of pest infestations. Conversely, where a single crop is grown extensively over a broad area, insects, which find this crop to be a host and thereby damage it, become dominant and the population diversity diminishes. A case where insect diversity can be profitably measured relates to the leafhopper population in rice-growing areas. Diversity measurement is done using the standard proportional abundance parameter:

$S = -\Sigma N_i \ln N_i$ where $N_i$ is the number of each species of leafhopper that is counted in a collected sample.

## Supporting activities that strengthen the relations between agriculture and the surrounding environment

A wide range of activities subsidiary to agronomy contribute to fulfilling concepts of compatibility. If desired, these can also be used for the construction of response indicators that apply to a particular setting. All of these activities take the form of practices that establish positive synergies between various essential activities that take place in many rural settings. Some examples of these are:

- Use of fallow land for animal grazing. In some places during a season when crops are not grown, land is leased out for a limited time for use by sheep and goat herders. The animals graze on the plants that have come up on their own, while the manure that they deposit contributes to soil fertility in the coming cropping season. The money derived from the operation also contributes to the durability of the farm enterprise.
- In tropical areas where there is a plentiful water supply, i.e. water in excess of the basic requirements for rice, paddy fields can be kept in a flooded condition with fish or shrimp being grown during the two months or so that is available. Again, an additional food source is produced and the waste generated by the fish contributes to soil fertility.
- A related activity practiced in Bangladesh and elsewhere is to raise chickens for eggs and meat in wire cages suspended over the paddy field. Again, droppings from the birds contributes to improved fertility in the plots.

## 212  Agricultural Sustainability

- The issue of use of 'waste' biomass has been discussed extensively in the section on efficiency. There are important compatibility aspects to this subject as well. Efforts to make use of any animal and plant waste (for which there is no higher value) for improving the soil closes production/consumption cycles within the area. Growing vegetables in a kitchen garden, composting the secondary material and then returning the compost to the soil is an especially commendable way to strengthen connections between humans and the rest of the environment.

### Box 3.6
### Strategy for assessing compatibility

- The scope of the assessment should be determined (area and time to be covered). For some parameters, compatibility measurements cannot be made on a single field or farm but must encompass a relatively broad area.
- Determine as many indicators as possible from those suggested for measuring human health issues (Ch1 to Ch5) and biodiversity (Cb1 to Cb3).
- Scale each of the chosen indicators to a common scale.
- Calculate the average within the two sub-category values, therefore providing an independent assessment of how the agricultural practices relate to human health and to the overall natural environment.
- The two sub-indices can then be averaged to produce a single compatibility index.
- Where appropriate, calculate response indicators for compatibility. Ch4 is one such indicator that among others suggests ancillary agricultural practices that support compatibility.
- Qualitatively evaluate the ways in which local practices are constructively compatible with cultural norms.
- Compare the water quality issues related to human health with those related to irrigation water in order to identify common problems, if they exist.
- Consider biodiversity issues as evaluated here. Note how the assessment affects understanding not only of compatibility but also of durability and productivity.

## 3.6 Equity

*Equity Agriculture should promote a good quality of life within families and among the various individuals involved in farming activities. This includes having consideration for the standard of living, health and education of all people in the community*

Yet another essential component of a sustainable agricultural system, beyond production of crops in an ecologically favourable manner, is the provision of a good measure of *social welfare* to individuals of both sexes, all ages and all social classes within the agroecological setting. All the aspects of agricultural activities must come together to support an adequate and comfortable standard of living. In essence, measuring equity is a means of assessing the state of human and social capital in the agroecosystem.

The term 'equity' is used here to signify a balanced distribution of the benefits of agriculture to all the members of the community. It is, of course, important to realise that the extent to which agriculture in any region has the potential to be productive depends on each specific environmental setting. Climate and soil are two of the defining features that limit this potential. It therefore cannot be within the realm of a sustainability study to expect that two areas—for instance, one shielded from rain by upwind hills and one exposed to plentiful and timely rains—be able to support the same variety of crops and level of productivity. For this reason, the extent of material benefits that contribute to a good life will inevitably also differ from place to place. We are here thinking mostly of equity within a given area. Equity between communities in different agroecological regions is an important issue, but one that usually goes beyond assessment of sustainability at the micro level.

On a larger scale, however, regional inequities should not be neglected. There are important contributions that can be made in terms of enhancing the agricultural potential of areas with problematic environmental settings—e.g. making choices such as the provision of crop varieties that are more resilient with respect to environmental stresses. Until recently, plant breeding efforts were directed substantially toward crops for which an ideal package of inputs could be provided. In particular, the ideal included the availability of an assured

## 214 Agricultural Sustainability

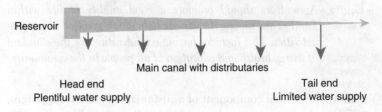

**Figure 3.10**
A canal irrigation system. Within the command area, more water is available at the head end near the reservoir, leading to distribution equity issues that require proper management strategies

supply of water. In more recent times, there has been at least a limited measure of research to develop improved varieties of crops (and the associated management practices) that are suitable for cultivation in arid and semi-arid regions. Some studies have also been focused on cropping systems appropriate for coping with other stresses, especially salinity. These efforts are very significant when one considers equity across broad regions.

The potential for creating community inequities within the command area of an irrigation project is another case in point. Problems of those farming at the head end taking advantage of their geographic location by extracting more than a reasonable share of water from the system are well documented. Often this leaves a deficit of water for those in the tail end, with resulting social and political tensions. Problems are also exacerbated when excessive water use upstream causes waterlogging and salinity problems downstream. In these situations, sound, enforceable management decisions, based also on good science that takes into account efficiency and environmental effects, are required to develop an equitable distribution protocol. These decisions should emanate from a balanced and fair consultative process that gives an equal voice to labourers as well as small and large landowners throughout the irrigation command area.

### Equity within an agroecosystem

In evaluating equity within a given agroecosystem, the principal concerns are distribution of profits among individuals living in the area as well as the choices that are made between supplying personal and

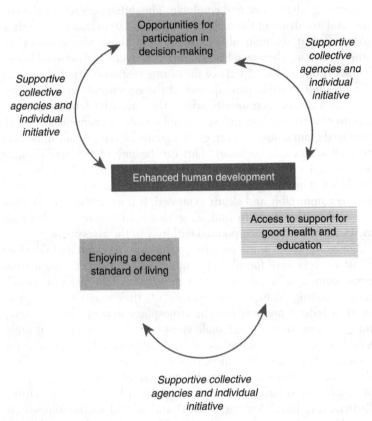

**Figure 3.11**
The key features of Enhanced Human Development (equity) based on the UNDP Human Development Report (2002)

individual profit versus supplying profit to the community at large. An equitable system is one that would provide employment for all healthy adults, would not create great disparities in the amount of land owned, and one that is beneficial to both landowners and hired labourers. An equitable system is also one that provides opportunities and benefits in a fair manner to females and males of all ages. Finally, an equitable system is one that makes available efficient public services, with the essential ones being services for education and health. Access to public services should be available to all, and decisions that affect the community should be made after broad public participation (see Figure 3.11).

With respect to each of these issues, there will be diverse opinions concerning what is just and equitable. The differences reflect the culture and worldview of the society and agroecosystem that is being studied. Different opinions also emanate from those who conduct the study, reflecting their own background, experience and personal biases. Equity is one of the aspects of measuring sustainability that requires the broadest possible participation of the community and the assessors. In a diverse community, where there may be tension between groups of high standing and wealth and socially or economically disadvantaged groups, and between gender groups, it is especially important that all voices can be heard. This can be one of the very difficult aspects of a sustainability study. The mere presence of representation for different groups, genders and ages does not ensure that all opinions can be comfortably and clearly expressed. It is incumbent upon those involved in the study to find some means to ensure that disparate views are heard and incorporated faithfully in the assessment.

For example, in some societies, extension, research and development workers have found that if one is to get 'correct' information from women, it is necessary that discussions take place in an all-female setting. Where men are present, they may be a source of overt or hidden pressure and the atmosphere may not be conducive for free expression of the female viewpoint. The same type of problem can arise in conversations with disadvantaged groups in some societies. The presence of more powerful and influential members of the same community has an inhibitory effect on the weaker members, who often have to serve as clients in a hierarchical situation. Participatory Rural Appraisal (PRA) and related methodologies are useful assessment strategies that provide opportunities for all members of the community to participate in studying, planning and implementation.

The United Nations balance sheet on human development (ESCAP 2002) lists a number of issues that should be considered in a detailed review of the status of social conditions in a given setting. We will consider those that are relevant to descriptions of sustainability in an agricultural community.

In the following sections, we examine issues surrounding social equity. Because there are great discrepancies between social situations within Low-Income Countries (LICs) and High-Income Countries

(HICs), we will document some important features regarding progress and deprivation in recent years for both. In that context, it becomes possible to develop simple indicators that allow us to evaluate the basic equity parameters. The issues chosen here are ones that have particular relevance to social situations in agricultural communities, especially those in low-income countries. Where available, other indicators that have been collected by various national organisations and non-governmental organisations can be used to provide supporting information that adds greater depth to the broad subject of social well-being.

The indicators suggested here will fall under the general headings of *educational opportunities, income and employment, gender fairness issues, support for human health,* and *food and nutrition.*

### Educational opportunities

There can be large differences between high- and low-income countries in educational facilities and opportunities for various levels and types of training. The importance of providing educational opportunities for all cannot be overemphasised. In the first place, this is a matter of social equity. Among other benefits, it is a means by which it becomes possible to train members of the rural community in practices that will enhance the development of sustainable agricultural practices where they live and around the world.

- In high-income countries, universal compulsory education supported by the state is the norm for children up to the age of approximately 16. Nevertheless, more than one-third of adults have attained less than upper secondary standing in education. Opportunities for tertiary education and training in different programs are widely available; financing ranges from full state support to a substantial portion being provided by the individual. Since 1960, the overall participation rate in tertiary education has risen from about 15 per cent to approximately 45 per cent in the first years of the new millennium.
- On the other hand, in low-income countries, broadly speaking, enrolment in primary education has increased from 48 per cent in 1960 to almost 80 per cent at present. Even in 2003, however, about 130 million children at the primary level and 275 million at

the secondary level are still out of school. The quality of this experience is highly variable between countries and often within countries. Government support for primary education is the norm but in some cases higher quality education is available through the private sector. In many situations, students leave school after a minimum number of years, typically about five. There are increasing opportunities to participate in tertiary education, but recent technological developments have widened the quality gap between HIC and LIC institutions. In many countries, much of the cost for tertiary education must be borne by the individual participant.

In addition to broadly based national data, local information about education is relatively easy to obtain in most circumstances and can be used to create micro-level indicators.

The following is a suggested state indicator to measure the current status of the local population in terms of level of education attained.

### Eqe1 (Educational status of the total population):

| | Individual score |
|---|---|
| Illiterate—no schooling | 0 |
| Primary schooling 0.5 points for each year up to 4 | 0.5 |
| Secondary schooling, 1 point for each year from 5 to 12 | 1 |
| Tertiary education, 2 points for completing a tertiary program | 2 |

The sum of individual scores is tabulated and an average value for the community calculated.

$$Eqe1s = (\Sigma x_i N_i / N_t) \times 10,$$

where $N_i$ individuals have a score of $x_i$, and the total number in the survey is $N_t$.

In the present indicator system a maximum score of 10 is achieved if everyone in the population has completed 12 years of education. Tertiary education then becomes a surplus benefit to the individual as well as to society.

The availability of education to the present generation, and the prospects for the future are measured by current enrolment in primary schooling; this leads to development of a response indicator.

**Eqe2 (Enrolment in School):** The percentage of primary school age children (ages 6 to 12) now enrolled in and attending school (P%) is determined.

$$Eqe2s = P\%/10$$

As noted earlier, the United Nations Human Development Index (HDI) uses an enrolment index as part of its education component, although that index measures a combined participation rate in the three levels of education.

## Income and employment

In recent years, macroeconomic and political policies followed in most countries throughout the world encourage global interdependence through trade. As a consequence, individual nations are less self-sufficient, instead relying on exports to support their own economies as well using imports as the source of many essential commodities. To some degree at least, persons in even the most remote agricultural communities participate in the global economy and require an adequate income in order to have access to materials required for a reasonable standard of living. Added to this are issues of income equity between various groups and individuals within the community. Sustainability demands that these subjects be addressed and resolved in order that humans live comfortably within the agroecosystem. There are marked differences in the current income and employment situation between high- and low-income countries.

- The situation in high-income countries is characterised by wealth production, but not always in a way that can be described as equitable. During the years from 1980 onwards, a period characterised by low inflation, real per capita income has grown by more than 2 per cent per annum, Wealth distribution is becoming increasingly skewed, with the wealthiest 20 per cent of the population getting around 40 per cent of the total national income, while the poorest 20 per cent get less than 10 per cent of the total. Total unemployment throughout much of the high-income world hovers at more than 8 per cent, with much higher levels experienced by youth in several countries and regions.

- Low-income countries too have undergone strong growth in many areas, but it has been uneven and often much of the benefit of wealth creation has been acquired by privileged groups, especially in urban settings.

Unemployment and underemployment are endemic throughout much of the low-income world, including in rural areas. During the years from 1980 onwards, real per capita income increased unevenly, averaging about 4 per cent per annum overall, with strongest growth in Asia and weakest performance in sub-Saharan Africa.

Over one-third of the total population exists with personal incomes placing them below the poverty level. There are great disparities between rich and poor. In rural areas, the disparities are often centred around access to land, with the landless poor being a particularly disadvantaged group. The distinctions between urban and rural poverty are noteworthy. In urban situations, the population often has greater access to organised physical resources such as safe drinking water, sanitation and health care. On the other hand, informal networks in rural societies are frequently a means of support for those whose financial resources place them below the poverty level.

Depending on the situation, it may or may not be possible to obtain the kind of monetary information needed to assess economic status and prospects. Essentially it is necessary to know about income and capital resources of individuals within the agroecosystem. Under the Productivity category of indicators, individual and regional income from crop production was the major consideration. There may however be other income components that should be added in order to give the overall farm income. In the case of agricultural labourers, daily wages may be the only source. Average income values as well as distribution among the population are both important components required to measure human development in the system.

For the following indicators, in most cases the most appropriate monetary unit to use is the local currency. This allows for comparison between regions within a country as well as comparisons with nationally defined values such as the poverty level. Where global assessments are being made, use of an international currency such as the US dollar or the Euro may become more appropriate.

**Eqi1 (Total Farm Income):** Total farm income ($I_{ti}$) for an individual farmer is the total net income from all crops and other sources ($\Sigma I_{1i} + I_{2i} + ...$) on each farmer's land over the annual cycle. Individual incomes can be summed for all farmers within a community or agroecoregion. Scaling requires selecting an optimum value (B) and a poorest value (P); the latter might be the local or national designated poverty level. For a region,

$$Eqi1s = 10 \times \Sigma(I_{ti} - P)/(B - P)N$$

(where N is the number of farmers surveyed in each area).

**Eqi2 (Land Value)** In most parts of the world, the value of land owned by farmers makes up a large portion of their personal worth. For an agroecosystem, average land value is the average monetary value of the land holdings of farmers within the region.

$$Eqi2 = \Sigma A_i \times V_i / N$$

where $A_i$ is the area (hectares) of land having value $V_i$ (value in local currency per hectare) in a given region and N is the number of farmers surveyed. Scaling would not usually be applied; this indicator would be most appropriately used for comparisons between regions within a particular country or area. It therefore becomes an indicator of interregional equity. The concept of best and poorest values is probably also irrelevant in most cases.

**Eqi3 (Farm Assets Value):** The other component contributing to the net worth of a farm family is the total capital assets (consisting of buildings, machinery, and livestock), and this can be calculated as an average for farmers within a given agroecosystem.

$$Eqi3 = \Sigma CA_i / N$$

(where $CA_i$ is the value of capital assets of an individual farmer in a given region and N is the number of farmers surveyed. As above, scaling is not appropriate in most cases.)

**Eqi4 (Income Equity Between Agroecosystems):** To establish a measure of equity in comparing two agroecosystems, e.g., an irrigated and a dryland area adjacent to each other, the average income of all farmers in the two regions is first determined. The ratio (R) between the two averages is taken, always using the larger number as the numerator. Choosing as best and poorest R values 1 (identical average incomes) and

10 (10-fold difference in incomes) respectively, the indicator calculation then becomes

$$Eqi4s = 10 - 10 \times (R - 1)/9$$

**Eqi5 (Income Equity Within an Agroecosystem):** A different income equity indicator describes income equity within a region of common agricultural practices, i.e., a single agroecosystem. To calculate this measure of equity, the average incomes of the top 50 per cent and the bottom 50 per cent (or some other percentages) of farmers are calculated. The ratio (R) is then determined as in the previous case. In order to scale the indicator, giving Eqi5s, the following set of values is recommended:

| Ratio | Score |
|---|---|
| <1.5 | 10 |
| 1.5–1.8 | 9 |
| 1.8–2.1 | 8 |
| 2.1–2.5 | 7 |
| 2.5–3.0 | 6 |
| 3–4 | 5 |
| 4–5 | 4 |
| 5–7 | 3 |
| 7–9 | 2 |
| 9–11 | 1 |
| >11 | 0 |

To illustrate how equity information can be used, data taken from the four well-defined agroecosystems in northern Karnataka State in India are again considered. The head end of the Tungabhadra Project Canal-irrigated command area is an area of high-intensity agriculture, mostly centred around rice. The tail end receives a less predictable water allotment and supports agriculture of a number of lightly irrigated crops. Outside the command area, the rainfed region relies on rather variable rainfall, averaging about 600 mm a year. The ancient area has been supplied with irrigation for more than 600 years by canals built during the Vijayanagara Empire and is an area of diverse crops scattered among the rocky hills that break up the landscape.

Income data (based on farmer surveys) within these four agroecosystems are shown in Table 3.16.

Table 3.16
Income distribution for farmers within and adjacent to the Tungabhadra Project (Karnataka State, India) project area

| Agroecosystem | Average income (all farmers)/Rs | Average income (top 50%)/Rs | Average income (bottom 50%)/Rs |
|---|---|---|---|
| Head | 194,000 | 244,000 | 118,000 |
| Tail | 55,500 | 96,600 | 21,200 |
| Rainfed | 31,400 | 53,300 | 7,340 |
| Ancient | 167,000 | 247,000 | 106,000 |

Source: Data for 1999 from vanLoon et al. (2001b)

Equity *between* agroecosystems within the total area can be assessed by Eqi3s. In areas served by canal irrigation, this is an important concept that depends strongly on the availability of water. Availability depends on the yearly supply from the reservoir but is also strongly dependent on the organisation of water management.

In this example, equity between the head end and tail end regions is given an indicator value as defined by Eqi4s:

$$\text{Eqi4s} = 10 - 10 \times ((194{,}000/55{,}500) - 1)/6 = 5.8$$

On the other hand, the value of Eqi4s when comparing the head end farmers with those in the adjacent rainfed area is 1.4.

Equity *within* an agroecosystem is illustrated by the information from the head end of the command area, where the ratio $R = 2.1$, giving an indicator (Eqi5) value of 7. For the tail end, the rainfed and the ancient areas, Eqi5 results are 4, 2 and 7 respectively. In this instance, the areas with highest average incomes are also the internally most equitable. Lowest average income and lowest income equity score were found to be in the rainfed area.

**Eqi6 (Total Wage for Farm Labourers):** The previous five income indicators relate specifically to landowning farmers. In a number of situations around the world, many persons within the agricultural community are without land and make a living by working as hired (by day, month or year) labourers. Work may be continuous or sporadic; measuring total wage therefore requires knowledge of wage rate and period of time over which wages were earned. The average wage rate within an agroecosystem is calculated by combining the wage rates for men and women and including wages for all normal (not specialised) farming operations. Wages are reported as hourly, daily or monthly, depending on the common practice in the region. Scaling would be a complex process in some situations, as it would require estimating the amount of time for which the labourer is given compensation during each month or year.

## Gender fairness issues

The complexity of gender issues is itself a subject for extended discussion. Around the world, there are vastly differing views on the subject, and everywhere these are evolving and frequently undergoing revolutionary changes. While it is sometimes asserted that the local culture should be the only determinant of what is fair and not fair, we believe that for a sustainable rural society there are certain basic issues that can and should be measured.

- In considering sustainability related to gender issues in agriculture, a number of features are pertinent in high-income countries. There are equal opportunities for education for male and female students in rural settings. In the tertiary education system, women now make up more than 40 per cent of the population. Women can make choices regarding the size of their families. Women can choose to participate in all aspects of agriculture, although this profession is still dominated by men. Women are active participants in decision-making with respect to the scientific, economic and social issues related to agriculture. But problems persist in a number of situations. Wage rates for comparable types of work are often higher for men than for women. Problems at the community and personal level continue to exist.
- In low-income countries there is a different set of issues pertinent to sustainable agriculture. Opportunities for education for females are increasing, but in some situations, social and economic conditions are a strong influence in limiting the extent to which females are able to take advantage of these opportunities. Women are more and more involved in family planning decisions. The fertility rates have on average declined significantly in rural as well as in urban areas. In some countries however,

social and economic conditions encourage a rural society made up of families having many children. Women work in agriculture in different ways, usually in some specifically defined roles. In some situations, they work extensively in the field, as much as or more than men; they are almost always also the principal workers in the home. There can be a significant differential in wages paid to men and women for work in agriculture. Opportunities for women's participation in decision-making regarding rural affairs in the home and in the community are usually less than for men, but vary from place to place.

Key indicators focus on statistics that are directed specifically toward women, including those that make comparisons with the male half of the population.

**Eqg1 (Gender-based Wage Differential):** This indicator measures the difference (D), in percentage terms, between the daily wage rate for men and women in comparable activities associated with agriculture. The optimal situation is wage parity between the sexes, while the poorest value is more arbitrary. In this example, we choose a 30 per cent differential as the poorest value. The scaled indicator is:

$$Eqg1s = (30 - D)/3$$

**Eqg2 (Gender-based School Participation Differential):** This indicator compares participation over a range of years (e.g., school levels from 1 to 10, or age levels from 6 to 16) of male and female students in schools in the region where the assessment is being made. A more detailed study might involve assessments at several individual levels. The form of the indicator would be similar to that described for measuring wage differential, again requiring an arbitrary choice for the poorest value in the scale.

$$Eqg2s = (30 - D)/3$$

Relative gender contributions to decision-making in rural affairs are more difficult to measure. This is especially true within the family unit itself. Many studies in low-income countries have documented women's contribution to seed collection and maintenance, making the female contribution in support of biodiversity greater than that of men. Decisions about crop selection are frequently a shared activity. Decision-making regarding other aspects of agriculture, especially those involving major purchases of machinery and farm implements, is more likely to be male-dominated. These issues are difficult to assess in quantitative terms. More amenable to measurement is participation in formal rural governance activities. The relative numbers of men and women on village councils, water user networks and cooperative councils is a good measure of the degree to which gender equity obtains in a particular area.

**Eqg3 (Gender-based Governance Differential)**   Once again, this differential can initially be measured in percentage terms (D) and scaled as in Eqg1 and Eqg2, assuming that parity on governing bodies is the ideal.

$$Eqg3s = (30 - D)/3$$

## Support for human health

There are a number of background issues regarding human health and access to environmental and health services that pertain to equity:

- In most high-income countries medical facilities of high quality are widely available, but access is variable, ranging from dependence on individual private health insurance to excellent comprehensive public or public/private health plans. Coupled with this, healthy living conditions are supported by almost universal access to clean drinking water, and food supply whose safety is well regulated and monitored. There are occasional exceptions to the high standards and there are concerns about environmental issues that can affect human health. For one thing, the level of treatment of waste water is highly variable and there are growing concerns about the quality of air, especially in urban areas. Nevertheless, indicators related to human health are generally positive. Life expectancy averages more than 75 years in many countries and infant mortality rates are low, usually ranging from about 5 to 15 per 1,000 live births.
- Over the past four decades, access to safe water in low-income countries has increased from about 30 per cent of the world's population to 80 per cent, and there is greater availability of waste

water treatment, although it is mostly restricted to urban areas. In contrast to these improvements, the quality of urban air has declined significantly, with large cities such as Mexico City, Beijing and Delhi having major and growing problems. There are very limited data regarding rural air quality. While outdoor air in the countryside may continue to be of acceptable quality, except adjacent to major urban areas, there are specific areas of concern. One relates to uncontrolled and excessive spraying of pesticides, sometimes under unacceptable atmospheric conditions. Another major area of concern is indoor air, especially where cooking is done over poorly-ventilated open fires using inefficient combustion technologies. The quality of the food supply is variable with regulations regarding food safety being minimal or non-existent. In most countries there are only limited high quality health care facilities and access is usually determined by individual financial status. On the other hand, there may be wide-spread access to very basic care, sometimes including care based on traditional medical systems. As a result of the somewhat greater availability of health providers and medicines, life expectancy has improved in many countries; in some it is now within 10 years of that in HICs. There are significant exceptions, most notably in sub-Saharan Africa where in recent years AIDS has caused a catastrophic decline in life expectancy to values as low as 39 years in countries like Malawi. Infant mortality rates are also variable, but typical rates range from 20 to 150 per 1,000 live births.

Indicators regarding health in a micro-region must take into account the differing societal backgrounds. National statistics of varying quality are in some cases available to provide information on a number of social issues, including life expectancy, infant mortality rates, number of doctors per unit of population etc. Where these data relate to a specific agricultural region, they can be used. Especially in the LICs, however, such data are not normally available on the local scale, and surrogate information is therefore required.

As a substitute for infant mortality data, one could determine local availability of facilities for childbirth. The level of support during childbirth can be obtained by individual interviews with mothers within the local population. The support is then ranked according to individual circumstances. An example of support possibilities defined by birthing facilities and ranking follows:

### Eqh1 (Settings Where Births Occur)

| | Individual Score ($x_i$) |
|---|---|
| In hospital or clinic | 10 |
| At home with assistance from trained midwife | 8 |
| At home with assistance from untrained midwife | 4 |
| At home with assistance from family members | 1 |

$$Eqh1s = \Sigma x_i N_i / N_t$$

where $N_i$ births attain a score $x_i$ and these are summed and averaged for all births. Data for a given ecoregion should cover a substantial period of time, for example, a ten-year period. The total number of individuals is $N_t$.

In the suggested ranking, home births with assistance from a trained midwife were considered somewhat less acceptable than hospital births, but they may in some cases be considered of equal acceptability. Further refinements could be made on the basis of the level of training of the medical practitioners assisting in the birthing event.

### Eqh2 (Cooking Facilities)

As a measure of indoor air quality, a health issue especially relevant to rural women and young children, the type of cooking system can be assessed. Issues include type of fuel, location of cooking and level of ventilation. A ventilated system has a well-designed and well-maintained flue or chimney, while a poorly ventilated system makes use of windows or other air passages.

| | Score ($x_i$) |
|---|---|
| Electricity | 10 |
| Ventilated CNG/LPG system | 10 |
| Poorly ventilated CNG/LPG system | 7 |
| Ventilated liquid fuel (e.g., kerosene) system | 8 |
| Poorly ventilated liquid fuel system | 5 |
| Ventilated solid fuel (coal, charcoal, biomass) system | 6 |
| Poorly ventilated solid fuel system | 0 |

$$Eqh2s = \Sigma x_i N_i / N_t$$

where $N_i$ is the number of cooking units of types that are assigned a score $x_i$ and $N_t$ is the total number of cooking units.

Clearly, some arbitrary choices are made in setting up the ranking scale from 0 to 10. These are based on the authors' observations in various village communities, but could be adjusted, refined and expanded for other individual situations.

The selection of this indicator raises an important issue. Many studies have pointed to major health issues associated with food preparation in poorly ventilated situations; for this reason it warrants a low score in terms of sustainability. The problems are most severe when the fuel is biomass, and it is burned in low-efficiency devices. It is for this reason that the indicator is an important one in pointing to actual or potential health problems. Yet there are other categories of sustainability issues involved in the use of biomass as fuel. One is its availability—it is clearly an inconvenience and a burden (and frequently an equity issue as well) when obtaining fuel requires time and effort to travel long distances in search of it. On the other hand, the use of a readily available local resource—in some cases secondary (waste) material that is a by-product of agricultural production of a primary food product—is in some ways a highly sustainable operation. This is especially true when the secondary material has limited value as a soil amendment or animal fodder. Sunflower stalks, cotton sticks and other woody products are examples of such materials.

### Food availability and nutrition

Another important measure of health, and also of quality of life in general, is the availability of nutritious food to the population as a whole. Nutritional status also shows great variability across the globe, but there can be differences on the micro scale as well.

- There are certain general characteristics of food availability and dietary status of the population in high-income countries. The food supply is adequate, and a high quality and varied diet is available to the vast majority of the population. Nevertheless, pockets of poverty leading to malnutrition are found in every country. More widespread are problems associated with over-consumption,

leading to obesity, and malconsumption, causing increased incidence of problems such as heart disease.
- In low income countries, food production has continued to increase at a rate greater than that of population growth. People with adequate means are able to eat well, usually much better than the equivalent population in earlier generations. Yet large numbers of people suffer from malnutrition due to poor distribution of available food supplies and to lack of financial resources for purchasing food. Close to one billion people in the world are inadequately nourished, over 95 per cent of them in LICs. While adequacy of nutrition is usually measured in terms of energy content (calories) of the foods, other issues, such as protein and vitamin supply, can be equally important in terms of ensuring a diet that prevents disease and provides for good health. A detailed examination of the diets of many marginalised persons in LICs indicates major deficiencies in these respects.

It is usually not difficult to estimate the energy value of food for individuals in a local situation. In many farm households, the amount of the staple grain or root crop eaten per year will be known. For example, a family of four might consume 'six bags of rice in a year'. If each bag of rice weighs 75 kg (the norm in India), then, using an energy value of $1.23 \times 10^7$ joules per kg (Table 3.11), the average energy content per day for each person is:

$$\frac{75 \times 6 \times 3.7 \times 10^7}{365 \times 4} = 11.4 \text{ million joules per day}$$

Using a conversion factor of 4,180 joules per food calorie, the daily calorific content (from rice only) for that family is 11,400,000 / 4,180 = approximately 2,700 calories. Grain alone will not make up the total energy content, although in most cases it will dominate the calculation. Similar estimates for other components of the diet can be made and the total value will give a first estimate of the quality of the diet as a whole. An important cautionary note is that the distribution of food within the family will almost certainly be unequal and could in some cases be inequitable. In some societies, women take their meal after the male members of the family, a practice that often leads to women eating much less. It is not unusual that in these same societies women not only share in sometimes back-breaking fieldwork on an almost equal basis with men, but also carry out most of the housework before and after their agricultural contribution. As a result, while consuming a smaller measure of food energy, they do work requiring a greater amount of food energy.

Recognising the limitations just described, a primary estimate of nutritional status can be calculated.

### Eqn1 (Average Daily Caloric Intake per Person)

| | Score ($x_i$) |
|---|---|
| <1600 | 0 |
| 1600–1900 | 2 |
| 1900–2200 | 4 |
| 2200–2500 | 6 |
| 2500–2800 | 8 |
| >2800 | 10 |

These scores provide individual values used to calculate Eqn1s.

$$\text{Eqn1s} = \Sigma x_i N_i / N_t$$

For families, communities or agroecoregions made up of $N_t$ persons, a composite value is calculated, where $x_i$ refers to the individual score for each person $N_i$.

While difficult to measure in a simple assessment, food quality aspects of diet are as important as food quantity. Protein content (both amount and composition) is perhaps the single most important individual dietary aspect, but availability of a variety of fresh vegetables and fruits that can supply adequate quantities of the vitamins and minerals needed for good health is also vital. It should also be remembered that food values can be lost as modern highly productive agricultural practices replace those that are more traditional. There is an attraction towards grains like white rice and for various types of processed foodstuffs. When these replace the coarser whole grains and other less altered products of the field, some elements of nutrition may be lost, with no adequate replacement as a substitute.

A separate but also effective way in which to measure the integrative consequences of nutritional status is to determine the percentage of children under 5 who are underweight (by some defined amount) for their age class. For example, one could determine the percentage of those who are more than 15 per cent under the normal weight and, using 0 per cent and 50 per cent as the best and worst goalpost values, the indicator would take the following form:

$$Eqn2s = 10 - P\%/5$$

where P% is the percentage of underweight children in the population sample.

### Response indicators related to equity

Various types of response indicators can be used to document ways in which different societies have established formal and informal institutions and processes to strengthen the well-being of the community. Some of these indicators might measure:

- Establishment of local fair-price shops to provide basic foodstuffs; school- or community-based feeding programs (such as noon-hour lunch); food-for-work programs.
- Schools of various levels built within the communities; affirmative action programs to encourage school attendance by disadvantaged groups.
- Land reform measures; minimum wage standards.
- Reservations for women and other under-represented groups in local decision-making bodies, schools etc.
- Provision of hospitals, clinics and primary health-care centres; support for local health workers, 'barefoot doctors', nurses and other paramedics.
- Presence of functioning co-operatives that support agricultural production, distribution and/or marketing of products.

It is again worth noting that there are many possible indicators that can be used to describe the many components of social equity and well-being. The parameters listed here are ones specifically related to rural settings in low-income countries. Where appropriate and available, these should be supplemented by conventional data from other sources.

> **Box 3.7**
> **Strategy for assessing equity**
>
> - The scope of the assessment should be determined (area and time to be covered). For the equity category, measurements related to individual persons carry little meaning in terms of overall sustainability. A substantial sample of the local population should be chosen.
> - Determine as many indicators as possible from those suggested for measuring issues related to educational opportunities (Eqe1s and Eqe2s), income and employment (Eqi1s, Eqi4s and Eqi5s), gender fairness (Eqg3s to Eqg1s) support for human health (Eqh1s and Eqh2s) and food availability and nutrition (Eqn1s).
> - Scale each of the chosen indicators to a common scale.
> - Calculate the average within the four sub-category values, therefore providing an independent assessment of how the agricultural practices relate to the various aspects of well-being.
> - The four sub-indices can then be averaged to produce a single equity index.
> - Evaluate other indicators that may not be appropriately scaled, but can be used to provide useful locally specific supporting information. Where available, use locally or nationally developed indicators to add further substance to the assessment.
> - Where appropriate, develop and calculate response indicators for equity.
> - Note again the overlapping health issues that are evaluated in the compatibility category.
> - Perhaps more than in any other category, participatory data collection methods are essential here. Concern about being sensitive in dealing with disadvantaged groups is a prime requirement in order to ensure the validity of the information obtained.

## 3.7 Evaluation of Agricultural Sustainability Assessment Protocols

Throughout this book we have tried to show that the design of a strategy for assessment of agricultural sustainability within a micro-region requires careful and systematic thought. Planning begins with the need to develop a clear overview of what is meant by sustainability in a very broad sense. Arising from this, a series of categories of assessment is set up, taking into account all the components that are considered essential. Finally, within these categories one establishes a set of indicators that can be used to evaluate the sustainability status under each heading.

Depending on the particular agricultural situation, there could be a wide range of potential overviews, categories and indicators that are relevant for the study. The strategies suggested in this book may however be applicable to a number of situations. Using some of these or similar procedures, one would end up with perhaps 40 or 50 individual indicator results, all scaled from 0 to 10, covering 6 categories that are related to the single issue of agricultural sustainability. Having accumulated this set of indicators, the questions that must be asked are: 'What is to be done with the collection of numbers? How can they be used, individually or in combination, to provide a measure of sustainability?'

### Combining indicator values

In many situations, the first step in the process is to combine values for individual indicators within the categories. If the same numerical scale has been used in each case, a simple average is often an adequate means of aggregation. There may however be situations where taking a weighted average is more appropriate. Consider a case where there are two sub-issues within a particular category and both are considered to be of equal importance. An example of this is an assessment of stability, where there are a number of indicators describing the integrity of the soil resource and a number of indicators related to water. If both of these sub-issues are considered to be of equal importance, weighting of indicator values should be done to reflect this decision. Thus, for example, if there are 3 soil indicators (with values of 6, 7 and 4) as well as 6 water indicators (whose values are 3, 2, 5, 6, 3 and 5), the former could each be given a weight of 1 and the latter a weight of 0.5 in calculating the stability average.

The calculation is then

$$\text{Stability average} = \frac{\overbrace{6 + 7 + 4}^{\text{soil indicators}} + 0.5 \times \overbrace{(3 + 2 + 5 + 6 + 3 + 5)}^{\text{water indicators}}}{3 + 0.5 \times 6} = 4.8$$

In this way, equal importance is given to both the soil and water issues, even though a greater number of indicators was available for evaluating stability of the water resource.

Having completed the averaging process for individual indicators within each category, a set of six category values has been obtained. The six individual numbers provide an assessment of sustainability in each of the chosen categories.

One can then go further and combine the six category scores, probably by taking a simple average, to give a single overall sustainability value. This can be called the sustainability index.

There will be situations in which different value levels—individual indicators, category indicators or sustainability index—will be the ones that are most useful. If an overall assessment of an agroecosystem has been carried out in which all the steps are followed, we might start by looking at the sustainability index. Where the index value is clearly very high, there is an obvious conclusion that, in a comprehensive fashion, agricultural management in that situation is highly sustainable. A fuller picture could involve reference to category and individual scores to give examples of the sustainable practices and features. The opposite picture is evident when the overall sustainability index value is clearly low. Again, examples from the categories and individual indicators would then make clear the basis on which the low rating has been obtained.

A more problematic situation arises in cases where the sustainability index falls in a grey zone that is neither high nor low. This intermediate score could arise in many different ways; it could be a consequence of mostly middle-range values within all the categories, or it could be due to a series of good values being balanced by several poor ones, giving an overall average in the middle range. For such

**Figure 3.12**
Combining individual indicators into category indicators and then into a sustainability index. This diagram is a subset of the diagram shown in Figure 2.6

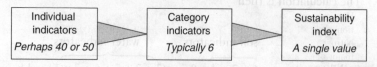

intermediate index situations, it is clearly not enough to be satisfied with saying that the assessment has given a value of, say, 5 out of 10. It is essential then to step back and examine the situation in each of the six categories. Perhaps it could be a case where high productivity is balanced by poor equity conditions. This raises its own set of issues. Many other combinations of category values could give a similar intermediate index and each one calls for a fuller explanation.

Just as the overall index is a product of several category values, each category indicator likewise is a product of a number of individual indicator values. For this reason, intermediate values within categories also require much deeper investigation. In the stability situation, for example, an intermediate score could arise from relatively good results for soil and poor ones for water. The source of problems or successes can only be determined by looking at each of the indicators.

The point is that, whatever values are obtained, indicators are merely numbers and they define the need to ask further questions. After all, it is the purpose of an indicator to point to successes and failures and to encourage us to investigate these in greater depth. Fortunately, the deeper investigation will already be under way, at least partially, through the processes of collecting data that were required to develop the indicators.

Having said this, the numbers can themselves serve a variety of useful purposes, as described earlier. Indicators and indices can be used to follow trends over time, for comparisons between similar situations in different places, to evaluate the progress of a project, and to assess different management practices in a given region. Publicising indicator results within the community can be an incentive for persons (individual farmers, farm managers, administrators, policy makers, etc.) to work toward making improvements—all in the service of striving for sustainable agriculture.

## Notes

1. Here and for subsequent indicator labels, a suffix 's' is provided where quantification and scaling methods are suggested.
2. Nitrogen content of plants is of interest also because this element is one of the major nutrients required for plant growth. Later we will show indicators that are used to measure the efficiency of added nitrogen uptake. To calculate these indicators, the nitrogen content of the plants is also required.
3. A statistically sound way of comparing variability is to carry out an F test. In this method, a ratio of the square of the standard deviations is calculated-in this case $0.732/0.292 = 6.3$. This number is set alongside a value obtained from a table of F values. In this instance, the comparison number is 4.3, and the greater value in the field results indicates that with 95 per cent confidence, the variability in the yields of Set 2 is significantly greater than that in Set 1.
4. This is calculated, assuming a bulk density of 1.3 tonnes per cubic metre, as $(1 \text{ t}/1.3 \text{ tm}^{-3}) \times 1/10,000 \text{ m}^2 = 8 \times 10^{-5} \text{m} = 0.08 \text{ mm} \approx 0.1 \text{ mm}$.
5. An excellent discussion of assessment methods used to indicate soil degradation, including erosion, is given in Stocking, M.A. and Murnaghan, N., Handbook for field assessment of land degradation, Earthscan Publications Ltd., London, 2001, 169 pp. The discussion presented here on soil quantity is based on descriptions in that book.
6. The process of organic matter degradation, which is an important soil forming activity, is thus a source of carbon dioxide release into the atmosphere. Clearly this contributes to the atmospheric burden of this greenhouse gas. Excessive degradation of organic matter is clearly undesirable, but degradation to produce a stable humus material should be encouraged. The organic matter-rich soil is then a more productive medium and the resulting enhanced growth of plants results in efficient sequestering of carbon dioxide into the new plant material. In this sense, the ability of microorganisms to increase the rate of carbon cycling in the local environment is considered to be an advantage.
7. A keystone species is one whose impacts on its community or ecosystem are large and greater than would be expected from its relative abundance within that ecosystem.
8. It is generally believed that evidence of the early stages of anthropogenic global warming will be highly variable yearly climatic patterns, often including local situations where there is excessive rainfall and others where there are major deficits, both occurring with increasing frequency.
9. One rating of photosynthetic efficiency of plants is: poor <0.06, range 0.06–0.12 normal, good >0.12.
10. Where large populations of livestock are kept in a small area, the disposal of waste can become a major problem. Use as a soil amendment, while intrinsically beneficial, becomes problematic when the ability of the soil to assimilate the animal waste is exceeded. Problems arise due to runoff and leaching of soil water. The two principal issues are loss of nutrients to receiving water bodies where they enhance rates of eutrophication, with detrimental affects to aquatic species like fish and their function as carriers of pathogenic organisms into water that finds use as a source of drinking water.

11  Methane, the combustible component of biogas, is also the principal component of natural gas.
12  In parts of KwaZulu Natal in South Africa, sugarcane is an important agricultural product. The cane is partially burned in the field prior to harvest in order to increase its sugar content, a practice that generates a large amount of particulates in the surrounding atmosphere. In some areas, like the city of Pietermaritzburg, which is located in a valley surrounded by hills supporting sugar plantations, a local atmospheric inversion develops. The inversion traps the cane-fire particulates near ground level leading to lengthy periods of degraded air quality for the 800,000 people living there.
13  The tragic increase in farmer suicides in parts of India, most notably Andhra Pradesh, in the early years of the new millennium is evidence of an agricultural system that lacks durability. In the early 1990s, farmers were encouraged to grow cotton and this was initially a highly lucrative enterprise because of high prices on the world market. Within the decade, however, market forces changed, and the world cotton price plummeted. Persons who had taken large loans to purchase inputs found themselves unable to pay back their debts. Several years of low rainfall caused a spiraling effect of increasing debt—a situation that led to great despair and ultimately suicide for many cultivators.
14  A much larger number of categories are included in the full set of guidelines (WHO 2003).
15  Data for this project were obtained by the authors between 1998 and 2000. A description is provided in Thompson, B., Hugar, L.B., Patil, S.G., and vanLoon, G.W., *'Cropping Patterns in Three Agroecosystems in the Tungabhadra Project Area of Karnataka State, India'*. A summary of sustainability issues in agroecosystems within the areas is given in Section 4.7.

# 4 Studies of Sustainable Agriculture

It is instructive to examine some of the many studies that have been carried out throughout the world with a view to measuring sustainability of agricultural practices. Of importance in the present context are not so much the results obtained in the studies, but the methodology used and the reasons provided for selecting particular methods. Each study is valuable in its own right; in most cases they focus on a limited number of aspects of sustainability, but there has been a wide variety of creative approaches used in selecting categories and specific issues, and also in assigning values and working out systems of aggregation. Consideration of the various approaches that have been developed and recommended can be very helpful when one is planning an assessment of agricultural sustainability in a specific situation.

## 4.1 A Study of Sustainability and Peasant Farming Systems in Zimbabwe

In this study (Campbell et al. 1997), the authors identify three key features that are essential for the analysis of sustainability: indicators of performance and the spatial and temporal scales of measurement. They recognise that selection of indicators is a value-laden process and choose to select three simple biophysical indicators.

The principal feature of the agricultural system in Zimbabwe is its distinctly dual nature, with half the land (in 1997) being held in large-scale highly productive farming operations owned by a single landowner and with much of the work being done by contract labour. The other half of the land (about 200,000 km$^2$)—a proportion that is growing—is taken up by peasant farming practices, typically with households each having access to around 3 ha of arable land. Maize is the major crop, with a mix of production directed toward direct consumption by the producers as well as some going to the market. Other crops include millets, groundnuts, sunflower and cotton. Besides cultivation of these crops, food variety and security is achieved by maintaining cattle, goats or chickens, and through harvesting activities related to the surrounding natural resources. Most households reserve a small area, typically where there is an assured water supply, for cultivation of vegetables.

Agriculture in the small farm areas relies on only a small measure of imported inputs, and water (rainfall) and low soil fertility are the two limiting factors in terms of productivity. The labour input is usually high. All family members may be involved, but much of the labour is done by women.

As measures of sustainability, the authors choose three indicators, each of which falls in the stability category. These are soil organic matter, soil erosion and crop yield. In each case, they emphasise the technical problems and limitations associated with using these indicators.

Soil organic matter levels in agricultural areas are typically lower than in the adjacent forest lands, and the organic matter including microbial biomass content drops rapidly and significantly when woodland is cleared in order to grow field crops. Levels of soil organic matter in cultivated fields are often below 0.3 per cent and the authors state that there is no clear fundamental or practical basis for assigning a target value. While the effect on soil organic matter of converting forests to agriculture is noted, other issues related to this practice are not considered in their assessment of sustainability.

In evaluating soil erosion indicators, the authors question the use of soil erosion data that are obtained from large-scale mechanised farming systems on clay soils. Such data estimate soil losses to be 50 to 75 tonnes per hectare per year, while soil formation is occurring at a rate of about 1 tonne per hectare per year. However, this would appear to vastly overestimate the problem in the peasant farming areas, where sandy soils predominate. In these situations, studies have measured soil losses as low as 2.4 tonnes per hectare per year, and reaching the higher range indicated above only during extraordinary situations.

The authors also question the use of sequential measurements of crop yield as an indication of sustainability. During the course of the study in the peasant farming regions of Zimbabwe, yields were found to be variable but increasing, while soil quality as determined by

**Table 4.1**
Some effects and responses of farmers during and after drought in small-scale farming systems in Zimbabwe

| During drought | After drought |
|---|---|
| *Livestock* | |
| Move or sell livestock | Reduce cattle loaning |
| Lop or prune trees for fodder | Rely more on donkeys and goats |
| Concentrate cattle on key resources | Rebuild herds |
| High livestock mortality | |
| *Crops* | |
| Expand wetland use for crops | Receive extra fertiliser |
| Reduce planting area | Plant later due to fewer cattle |
| Reduce fertiliser use | Use extra fertiliser saved from drought |
| | Use less manure due to fewer cattle |
| | Switch to small grains |
| *Household* | |
| Seek wage employment | |
| Increase craft work | |
| Market wild foods and wood | |
| Rely more on purchased food | |
| Reduce purchases of luxury goods | |
| Rely on emergency aid | |

**Source:** Summarised from Campbell et al. (1997).

organic matter and erosion data was evidently declining. The increased yields were associated with improved seeds and the growing use of inorganic fertilisers.

In the Zimbabwe study, issues of space and time are dealt with in some detail. The great variations that are observed over a range of family farms, all of which are managed in heterogeneous ways, means that individual observations should be extrapolated to a larger scale only with great care. The importance of considering the broader sociopolitical framework is also emphasised. Likewise, when considering the time scale, external forces including 'villagization', climate and changes in technology and other social and economic factors means that assessments of sustainability could only be considered to be realistic as a predictor extending to a period of 100 years or less. The role of diverse activities in coping with stresses such as drought is discussed. Words like durability or resilience are not used, but a number of important stress response factors and activities are noted (see Table 4.1).

The study points out problems that can arise when a superficial treatment is used in measuring sustainability and emphasises that there are limitations inherent in each indicator. Difficulties in aggregating information from individual farms to provide more broadly applicable data are pointed out, but this also supports the need to obtain data at a variety of levels. It is the interplay of both types of information that allows for development of the general picture and policies that would support sustainability in a region, while also taking into account variations and individual needs. The importance of going beyond the biophysical factors of sustainability to include social and economic issues is also emphasised.

## 4.2 A Malaysian Study of Farmer Sustainability

Taylor et al. (1993) have developed a composite Farmer Sustainability Index (FSI) that is designed to measure the degree of sustainability of individual farm management practices followed in the production of cabbage in Malaysia. They recognise that there are no universal principles for defining an absolutely sustainable system, but there are a variety of practices that, taken together, can express the degree of sustainability of individual operations. The emphasis is on biophysical processes that support sustainability, particularly those that replace imported inputs with ones that are available on-farm. Examples of on-farm resources include:

- Use of integrated pest management for controlling insects
- Employing crop rotations, intercropping and relay cropping to enhance soil fertility, control weeds, and maximise use of space and time
- Soil incorporation of livestock wastes, crop residues and green manures to enhance fertility and good physical properties
- Promoting nitrogen use efficiency by sequestering it from the air and making it available to crops through nitrogen-fixing legumes
- Encouraging mineral release and recycling from soil reserves
- Making water available to crops by applying water harvesting and other enhanced soil moisture retention strategies

**Table 4.2**
Ratings used to assess practices followed to maintain and enhance soil fertility in cabbage production in Malaysia. Score values can range from −5 to +13, giving an overall distribution of 18 points

| Type of production practice | Indicator score |
|---|---|
| Per cent of total applied nitrogen from organic sources | 0% = −1; 1–20% = 0; 21–40% = +1<br>41–60% = +2; 61–80% = +3<br>81–99% = +4; 100% = +5 |
| Number of split applications of inorganic fertiliser | 0 or 1 applications = 0;<br>2 applications = +1;<br>3 applications = +2 |
| *Changes over last five years in application* | |
| Inorganic fertilisers | more now = −2; no change = 0;<br>less now = +2 |
| Livestock manure | more now = +1; no change = 0;<br>less now = −1 |
| Other organic fertilisers | more now = +1; no change = 0;<br>less now = −1 |
| *Other means to enhance soil fertility and health* | |
| Follow crop rotations with legumes | yes = +1; no = 0 |
| Use slaked lime | yes = +1; no = 0 |
| Use paddy husk carbon | yes = +1; no = 0 |

*Source*: Taylor et al. (1993).

- Selecting crop varieties on the basis of their resistance or tolerance of insects and diseases
- Construction of bunds and terraces to control soil erosion
- Modifying planting dates and other cultural practice
- Control of management and labour by the farm family

Various agricultural practices were evaluated in an attempt to measure sustainability. These included practices for control of insects, disease and weeds, and practices to control erosion and to maintain and enhance soil fertility. Scores were obtained within each of five categories and summed to give an overall rating. In aggregating the quantitative values obtained within and among categories to come up with a single FSI score, items were assigned weights according to how strongly each practice was believed to influence sustainability. Scoring within the soil fertility category illustrates the nature of the weighting system (Table 4.2).

In assigning weights to practices used in soil fertility control, the overall issues considered were the amount of fertiliser purchased and used (although this was not measured in an explicitly quantitative way), the ratio of inorganic to organic fertiliser applied, the number of applications made during the growing season and the types of non chemical practices followed. In general, organic sources of nutrients (particularly nitrogen) were considered to be most desirable from the point of view of sustainability, and the largest weighting (+ 4 to −1) in the fertility category was given to this issue. Best score was assigned when most but not all nitrogen was obtained from organic sources, A somewhat lower score was assigned when a farmer used organic fertilisers exclusively. This was because these sources may not supply nutrients in exactly the proportion required by crops and soil.

In aggregating the different categories, 53 per cent of the weight was assigned to insect control, 22 per cent to soil fertility, 11 per cent to disease control, 7 per cent to weed control and 7 per cent to soil erosion control and other practices. The weightings were determined using an internal validation procedure involving measuring the correlation between individual measurements and the final FSI, as well as by an external validation through consultation with a panel of scientists.

Employing this system, a wide range of sustainability practices were considered together in order to establish a comprehensive index. Most of the issues fall within categories related to stability and durability of agricultural practices. Also recognised in the study was the concern of producers and policymakers that sustainable farming should not result in lower yields and profits compared with conventional higher-input agriculture. However, productivity was not included in the calculations, and would have to be considered as a separate issue that is determined in a parallel but independent evaluation.

## 4.3 Sustainable Land Management in Vietnam, Indonesia and Thailand

This study (Lefroy et al. 1999) focuses on issues of sustainable land management, especially in slopelands in South-East Asia, and begins with the premise that land management involves issues that go beyond the subject of soil quality. Sustainable land management 'encompasses the need for long term preservation of the resource base to allow adequate future food production in a manner that is socially acceptable, economically viable and environmentally sound.' In this definition are included aspects of the three components of sustainability (environment, economy, society). Indeed, the Framework for Evaluating Sustainable Land Management (FESLM) incorporates assessment of each of these components.

Using a participatory process involving expert sources as well as information from members of the farming community, 'five pillars of sustainability' were established in the FESLM protocol and included *productivity* (maintenance or enhancement of production services), *security* (reduction in the level of production risk), *protection* (protection of the quality of natural resources and prevention of degradation of soil and water quality), *viability* (economic security and equity) *and acceptability* (serves social needs). These recommended categories, while differently arranged and named, relate closely to those that have been proposed and described in this book. Productivity is broadened to include issues that we have called stability, security is essentially synonymous with durability, protection contains some elements of compatibility, and viability and acceptability together provide a comprehensive assessment of issues of equity.

Indicators selected in the prototype FESLM methodology are listed in Table 4.3.

The data were evaluated and aggregated using a scoring and ranking system whereby indicators were scored as being strategic, cumulative or suggestive, with weights of 10, 7 and 3 respectively. A variety of ranks were assigned, with relative ranks from 1 to 10. The value of the indicator was then taken to be the product of the score times the rank.

Once individual indicator values were calculated, they were combined within each category and rated on a scale as follows:

- 1   practices meet sustainability requirements
- 2   sustainability is marginally above the threshold
- 3   sustainability is marginally below the threshold
- 4   practices do not meet sustainability requirements

Using the FESLM methodology, 20 farms in Thailand were assessed; half were project sustainable farms and half were selected from

**Table 4.3**
Indicators developed in the Framework for Establishing Sustainable Land Management (FESLM) Project in South-East Asia

| Indicator | Qualitative, semi-quantitative and quantitative ratings |
|---|---|
| *Productivity* | |
| Yield (average of 7 years) | > village average |
| | < village average by 0–25% |
| | < village average by ~25% |
| | < village average by >25% |
| Soil colour | dark soil (high organic matter) |
| | brown soil (medium organic matter) |
| | yellowish soil (low organic matter) |
| Plant growth | vigorous |
| | normal |
| | stunted |
| Leaf colour | dark green |
| | normal |
| | yellowish on whole leaf |
| | yellowish on tips and margins |
| | older leaves purple |
| *Security* | |
| Average annual rainfall | excessive > 2400 mm |
| | sufficient = 1200–2400 mm |
| | limited < 1200 |
| Residue management (Percentage returned) | 50% for 3 years or more |
| | 50% for less than 3 years |
| | < 50% for 3 years or more |
| | < 50% for less than 3 years |
| | burnt or removed |
| Drought frequency | > 2 years continuously |
| | 2 years in 7 |
| | < 2 years in 7 |
| Income from livestock | > 25% of total income |
| | 10–25% of total income |
| | < 10% of total income |
| *Protection* | |
| Topsoil eroded (amount in last 7 years) | > 4.5 cm lost, rills on > 50% |
| | 0.7–4.5 cm, rills on 25–50% |
| | < 0.7 cm, rills on < 25% |

*(Continued)*

*(Continued)*

| | |
|---|---|
| Cropping intensity and extent of protection | 2–3 crops with conservation<br>2–3 crops with no conservation<br>1 crop with conservation<br>1 crop with no conservation |
| Cropping pattern | rice or corn, then fallow<br>rice, then corn<br>rice/corn, then legume<br>rice/corn between perennial |

*Viability*

| | |
|---|---|
| Net farm income | rising (B:C* > 1.25)<br>constant (B:C = 1)<br>declining (B:C < 1)<br>fluctuating |
| Off-farm income | > 25% of total income<br>10–25% of total income<br>< 10% of total income |
| Difference between market and farm price | > 50%<br>25–50%<br>< 25% |
| Availability of farm labour | 2 full-time adults<br>1–2 full-time adults<br>1 full-time adult |
| Land holding size | < 1 ha<br>1–2 ha<br>> 2 ha |
| Availability of farm credit | > 50% of requirement<br>25–50% of requirement<br>< 25% of requirement<br>not available |
| Percentage of farm produce sold in market | > 50% sold<br>25–50% sold<br>< 25% sold |

*Accessibility*

| | |
|---|---|
| Tenurial status | full ownership<br>long-term user rights<br>no official land title |

*(Continued)*

*(Continued)*

| | |
|---|---|
| Access to extension services | full technical support<br>limited technical support<br>no support |
| Access to primary schools | < 1 km<br>1–3 km<br>> 3 km |
| Access to health center | < 3 km<br>3–5 km<br>> 5 km |
| Access to agricultural inputs | < 5 km<br>5–10 km<br>> 10 km |
| Subsidy for conservation practices | > 50%<br>25–50%<br>< 25%<br>no subsidy |
| Training in conservation practices | once in 3 years<br>once in 5 years<br>not available |
| Village road links to major roads | full access<br>limited access<br>no access |

\* B:C = before : current
Source: Lefroy et al. (1999).

surrounding non-project participants. In the case of the project farms, all met or were well above threshold values for sustainability in all except the security category, while the non-project farms showed themselves to be far less sustainable.

The information obtained in the FESLM studies was developed for use by extension workers and staff of non-governmental organisations as a means of categorising farm management practices as being sustainable or non-sustainable, and therefore providing means by which measures could be taken to develop innovative practices toward achieving sustainability. The study involved methodology development, but other examples of its application were not provided.

| 4.4 | **An Integrated Crop Management Approach to Sustainable Agriculture** |
|---|---|

In response to the intensification of agriculture in low-income countries as a means of producing more food and other agricultural products, Meerman et al. (1996) have described an integrated crop management system that brings together a need for high productivity with techniques of sustainability. Their recommendations are limited largely to the biophysical aspects of agriculture, and include recommendations for land reclamation, soil and water management, cropping practices, pest management and use of plant genetic resources.

Recognising that if one is to put sustainability into practice, there is a need to develop methods by which agro-ecological sustainability can be measured, they suggest 10 characteristics to be monitored (see Table 4.4).

Using these or related kinds of indicators, data were collected to monitor a variety of types of agricultural practices, mostly in Europe and North America. The commonly-observed findings include the observation that pesticide use could be eliminated or reduced dramatically compared to conventional farm methods. In some cases, this resulted in a small (< 20%) decrease in yields, but in these cases profit margins could be maintained because of the reduced need for chemical inputs.

Meerman et al. emphasise the need to be flexible in developing a monitoring system. Not every sustainability parameter that was defined by them needs to be assessed, and the frequency of assessment also depends on the specific indicator and the local situation. For marginal soils and subsistence agriculture, soil physical and chemical properties might be most important, while for cash crops such as vegetables and cotton, the significant issues would more likely be associated with non-renewable inputs, including chemical fertilisers and pesticides.

Being flexible also means being open to the development of new indicators, applicable in specific situations. This is a key feature in the development of biological indicators that can detect short- and long-term pest and plant-stress problems.

**Table 4.4**
Criteria for monitoring agroecological sustainability

| Property measured | Measurement criteria for sustainability |
|---|---|
| Nutrient balance sheet | Nitrogen input/output ratios are measured, taking into account all inorganic and organic forms, including those from inorganic, animal, plant and atmospheric sources. A ratio of 1 is considered to be optimal. |
| Vegetation cover and composition | Extent and composition of soil cover are determined. Increased cover (for water retention) and diversity of species are considered desirable. |
| Water infiltration and run-off | Measure infiltration capacity or bulk density. High infiltration/run-off ratio indicates sustainability. |
| Replenishment and use of fossil water | Various data can be used to measure ratio of replenishment/use, with high values indicative of sustainability. |
| Economic threshold for pest problems | Measured by pest numbers and crop damage, as well as by biological control methods. Sustainability high when threshold is not reached using non-chemical control. |
| Pest complex | Measured by distribution of pest species. Sustainability is high when there are minimal large-scale outbreaks of pest problems. |
| Host plant resistance | Origin and type of resistance in commercial varieties is measured. Sustainability is lower when there is an accelerated need to develop pest resistant varieties. |
| Pest resistance to pesticides | Level of pest resistance to pesticides is measured; sustainability decreases as resistance level increases and multiple resistance arises. |
| Biological control agents | Sustainability increases as biological control methods rather than chemical methods take an increasing share of pest control situations. |
| Pesticide use | Minimal application rate (number of sprayings, dosage rate, area sprayed, potency of pesticide) indicates greater sustainability. |

**Source**: Summarised from Meerman et al. (1996).

## 4.5 Indicators for Comparing Performance of Irrigated Agricultural Systems

Workers at the International Water Management Institute in Colombo, Sri Lanka, have developed a set of indicators specifically directed toward assessing the performance of irrigated agroecosystems. These indicators can serve the purpose of comparing performance in large command areas around the world, with a view to optimising conditions for sustained and high biological and financial productivity. Special emphasis is placed on the efficiency or water use. Molden et al. (1998) identify the following reasons for assessing performance: 'to improve system operations, to assess progress against strategic goals, as an integral part of performance-oriented management, to assess the general health of a system, to assess impacts of interventions, to diagnose constraints, to better understand determinants of performance and to compare the performance of a system with others or with the same system over time.' While sustainability is not explicitly mentioned in the list of reasons, given that water is nearly always a limited and limiting resource, performance (i.e. in a general sense maximising the efficiency of water use) of irrigation systems is intrinsically a sustainability issue.

The small set of indicators chosen are based on phenomena that are common to most irrigated systems and allow for use of easily and commonly obtained data. Four of the indicators relate to productivity and productivity efficiency in terms of water use:

1. Output per cropped unit area = $\dfrac{\text{production (in monetary units/season or year)}}{\text{irrigated cropped area (ha)}}$

Production is the output of the irrigated area in terms of gross or net value, measured in units of either local or global currency, depending on which comparisons might be made. The irrigated cropped area refers to the sum of cropped areas during the time period of the analysis, typically either a single cropping season or one year.

2. Output per unit command area = $\dfrac{\text{production (in monetary units/season or year)}}{\text{command area (ha)}}$

This indicator is similar to the first indicator, but takes into account the entire command area and therefore measures output for the system as a whole, also including inefficiencies or other problems associated with distribution of water throughout the entire area designated to be irrigated. This can then provide critical data to indicate poor water management factors, such as occurrence of salinity, that have led to reduced productivity in parts of the total system.

3. Output per unit irrigation supply = $\dfrac{\text{production (in monetary units/season or year)}}{\text{total irrigation supply } (m^3/\text{season or year})}$

The total irrigation supply is the volume of surface irrigation water diverted to the command area plus net removals from groundwater. The indicator is not a measure of total water use efficiency, since it does not take into account water obtained from rainfall.

4. Output per unit water consumed = $\dfrac{\text{production (in monetary units/season or year)}}{\text{volume of water consumed by ET } (m^3/\text{season or year})}$

The water consumed is obtained by subtracting, from the total irrigation supply, water losses from fallow land, from open water surfaces, weeds, trees, flows to drainage channels, and water not used because of pollution.

Aside from these four productivity indices, two (1 and 2) of which are measures of yield and the other two (3 and 4) are measures of efficiency of water use, Molden et al. also propose a Standardised Gross Value of Production (SGVP) index, which is applicable in a variety of situations, both irrigated and rainfed. The SGVP index is a monetary index that allows for comparisons between systems at different times and located around the world, even when there are variations in prices with time and differing values of local currencies. To calculate the SGVP, an equivalent value is calculated, based on the local prices of the crops grown, all expressed as a ratio with the local price of a predominant, locally-grown and internationally-traded base crop. This ratio is then multiplied by the standard international price of the base crop.

The formula for calculating this index, applicable in the Molden et al. report to an irrigation system command area, but also extendable to other agroecosystems, is as follows:

$$SGVP = (\Sigma A_i Y_i P_i / P_b) P_{world}$$

where $A_i$ is the area planted to crop I (ha), $Y_i$ is the yield of crop I (kg/ha), $P_i$ is the local price of crop I (local currency units), $P_b$ is the local price of the base crop (local currency units), and $P_{world}$ is the price of the base crop traded on the world market ($ US).

The SGVP is therefore 'normalised', so that any crop can be valued with respect to a common commodity trading in a given year at a set price. A case is made for using gross crop values rather than net values in this and other indicators where productivity is measured in currency units. There are two reasons for this. One is that net values are subject to a variety of distortions associated with difficulties in assigning costs to land, water and family labour as well as to variations in taxes and subsidies on inputs used in many situations.

Once the SGVP has been determined, it can be used as the measure of output in determining any one of indicators 1 to 4. As noted above, this allows for comparisons of irrigation efficiency in different ways between irrigated areas supporting various types of agriculture around the world.

Five other indicators are proposed that characterise the individual system with respect to water supply and finances. Two are basic water supply indicators.

$$5.\ \text{Relative water supply} = \frac{\text{total water supply } (m^3/ha)}{\text{crop demand } (m^3/ha)}$$

$$6.\ \text{Relative irrigation supply} = \frac{\text{irrigation supply } (m^3/ha)}{\text{irrigation demand } (m^3/ha)}$$

In the case of the relative water supply indicator, crop demand is defined by the potential evapotranspiration (ET) or the ET under optimal well-watered conditions. The ET value can be obtained from locally available data or determined using a programme such as CROPWAT (FAO 1998). For relative irrigation supply, irrigation demand is crop ET minus water supplied by rainfall. The irrigation water supply is that derived from diverted water plus groundwater sources, whereas the total supply also includes rainfall. Both these indicators relate supply to demand, and optimal use would normally be indicated by values of 1 or slightly less. The scale of measurement is important here. Consider a command area where the relative irrigation supply, calculated for the area as a whole is 1. If head end values are considerably greater than 1, leaving values at the tail end to be very small, this is an indication of inequitable distribution—as we have seen, this is a common problem in poorly regulated irrigation systems. For this reason, values that average a range of situations may mask problems on the smaller scale. As with many indicators, using data at a variety of scales (when available) often provides useful insights that are lost in indicators that cover a larger region.

$$7.\ \text{Water delivery capacity} = \frac{\text{cancel capacity to deliver water at the system head (m}^3/\text{month)}}{\text{peak consumptive demand (m}^3/\text{month)}}$$

Water delivery capacity is an indicator that measures the ability of a canal irrigation system to provide water under conditions of highest demand. The peak consumptive demand in the above formula is the irrigation requirement expressed as a flow rate at the head of the irrigation system.

There are two financial indicators that provide information about the long-term financial viability, and therefore the economic sustainability, of the system as a whole. Both can be expressed either as a fraction or percentage.

$$8.\ \text{Gross return on investment} = \frac{\text{SGVP (\$ US)}}{\text{cost of irrigation infrastructure (\$ US)}}$$

The cost of irrigation infrastructure refers to the delivery system and one can calculate the appropriate value for the period of time over which the productivity index is measured as a proportion of the total lifespan of the infrastructure. Although Molden et al. specify relative calculations with reference to international currencies, a similar index could be evaluated using total productivity and irrigation costs expressed in units of local currency.

9. Financial self sufficiency = $\dfrac{\text{revenue from irrigation (local currency/year)}}{\text{total expenditure on operation and maintenance (local currency/year)}}$

Revenue is made up largely from farmers' fees for water use; expenditures incorporate all operation and maintenance fees including those for administration.

These two indicators allow one to estimate the extent of benefits derived from the irrigation system and the degree to which the system is self-sustaining or, alternatively, the extent of subsidy required.

The recommended indicator set was tested in 18 individual situations in 11 countries in Africa, Central and South America and Asia.

## 4.6 Developing Indicators: Lessons Learned from Central America

This project in Central America (Segnestam 2000) was directed toward developing and using indicators in order to measure and track issues of sustainability in Central America. Emphasis was placed on the ability of a user-friendly set of indicators to provide a basis for policy development and action. A conscious decision was made to include the three components of sustainability, both as individual categories and in combination with each other.

Indicator development and monitoring was done at a range of levels. At the national or regional level, a small number of aggregated indices were proposed to allow for comparisons over time and between regions. The indices were calculated by combining a series of core indicators that had been assessed in each country. To create a more detailed picture of local situations, a larger number of complementary indicators were proposed; these were considered to be optional and could be combined in a flexible manner according to conditions in particular locations. In the development of the assessment program, a regional network for Central America was set up, with a system of consultation between representatives from the six countries. Consultation involved persons at relatively high levels in the policy-making process as well as workers in governmental and non-governmental institutions that deal with agriculture.

**Table 4.5**
Sustainability indicators related to agriculture in Central America

| Issue | Pressure | State | Impact | Response |
|---|---|---|---|---|
| a. Land use | 1. Land use change (%/y or ha/y) | 2. Production system (ha) | 3. Soil degradation (ha) | 4. Potential agricultural yields (J/ha or t/ha) |
| b. Forest | 5. Deforestation (%/y or ha/y or %/natural) | 6. Forest cover (ha/type) | 7. Forest fragmentation (% or ha) | 8. Action plan for forest management (yes or no) |
| c. Fresh water | 9. Water sectoral withdrawals (% or $m^3$) | 10. Water consumption per capita ($m^3$) | 11. Water availability per capita ($m^3$) | 12. Rural population with access to safe water (%) |
| d. Biodiversity | 13. Conservation condition (ha) | 14. Non-domesticated land (% or ha) | 15. Eco-regions needed for conservation (ha) | 16. Protected areas (% or ha) |
| e. Marine and coastal resources | 17. Population in coastal areas (number) | 18. Mangroves and corals surface (ha) | 19. Areas polluted in coastal zones (ha) | 20. Protected marine and coastal areas (ha) |
| f. Atmosphere | 21. Location of fires | 22. Greenhouse gas emissions (T of C, by activity) | 23. Greenhouse gas emissions (T of C) per capita | 24. Participation in treaties and conventions (yes or no) |
| g. Energy | 25. Energy consumption per capita (by source or total, J) | 26. Hydropower generation (total J) | 27. Efficiency of dams (kW/ha) | 28. Hydropower potential (J/y) |

| | | | |
|---|---|---|---|
| h. Social dynamic | 29. Distribution of population (rural, urban) | 30. Literacy rate (%) | 31. Population in poverty (%) | 32. Projected population change (number) |
| i. Economic dynamic | 33. Structure of production (%) | 34. Structure of employment (%) | 35. Unemployment rate (%) | 36. Structure of exports (%) |
| k. Infrastructure | 37. Road network (km) | 38. Population with access to sanitation services (%) | 39. Infrastructure distribution (electricity, dams, roads, hospitals, schools) | 40. GDP in infrastructure (%) |
| l. Natural events | 41. Frequency of natural disasters (number/y) | 42. Population affected by natural disasters (number) | 43. Economic and human loss due to natural disasters (value or number) | 44. Areas susceptible to natural disasters (ha) |

**Units of measurement**

J/ha   joules per hectare
t/ha   tonnes per hectare
J/y    joules per year
T of C tonnes of carbon
m³     cubic metres

**Source:** Summarised from Segnestam (2000).

#### Figure 4.1
A project-based framework for a study of sustainability of water supplies at the regional level in Central America. Collection of data, development of indicators to measure inputs and outputs lead to improvements in accessibility to water of appropriate quality (redrawn in modified form from Segnestam 2000).

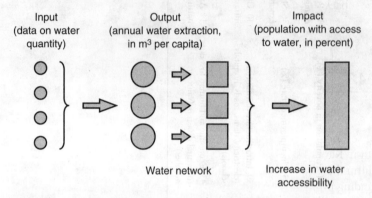

Within each issue for which information was required, four related indicators were proposed using a 'pressure, state, impact, response' approach. The list of 44 core indicators is presented in Table 4.5.

Besides obtaining data for core indicators over a range of regions—data that could be used for comparisons between areas and for creating a time series as a measure of progress—specific sectoral studies were also carried out. These were project-based and custom-made within a practical framework. The goals for the sectoral studies were to determine the kinds of data required, ensure data quality, develop indicators relevant to the diverse situations, and use the information gathered to stimulate real project outputs on the ground. For example, a framework for a study on water use was developed; the framework focused on water quantity issues, with an ultimate goal of increasing water accessibility (see Figure 4.1).

In evaluating the project, the researchers emphasised several factors:

- The need for a long-term strategic focus related to sustainable development in the obtaining of information
- A flexible focus and framework, in order to take into account different regions and changing situations
- The importance of using available data, ensuring its quality and modifying it to create a more complete picture of the situation

- The importance of developing ownership from policymakers, indicator developers and users of the information
- Designing the indicators and the way in which they are to be used with a view to supporting policy development
- Disseminating the information broadly so that it is used and feedback may be obtained, which can be used to sustain the project.

## 4.7 Sustainability in the Tungabhadra Project Area of South India

At several points in the book we have referred to research work carried out in Karnataka State of South India, directed towards assessing agricultural sustainability in and around an irrigation command area. A guiding principle of this assessment was the belief that a weak link in many studies (especially macro-studies) of sustainability is that they do not take into account the specific and often unique details of local situations. This is particularly true in a country like India, where spatial variations of language, culture, and biogeophysical characteristics are so distinct. For this reason, micro-sustainability measures are essential, important in their own right for describing features within a small region, as well as a basis for building up a broader picture of national sustainability. It was with this in mind that we set out to develop indicators that faithfully enable assessment of sustainability of agricultural activities in villages in and around the Tungabhadra Project (TBP) area in northern Karnataka, India.

### Sustainability of agriculture in the Tungabhadra Project area of Karnataka

In India, changes in agricultural practices are occurring rapidly and a variety of food and fibre production systems coexist in near proximity throughout the country. With its large population, India must assure that there is high productivity of all the essential crops, and to a great extent this has been achieved in the past 30 years. Yet, at the same time, food security can be guaranteed only if socially, economically and environmentally sustainable agricultural practices are followed.

The assessment attempted to measure, in a holistic manner, issues of sustainability of the agricultural systems used in and around the Tungabhadra Project area of South India. The TBP is situated in a semi-arid region of northern Karnataka State and Andhra Pradesh. It is the first major irrigation project completed in independent India and serves a command area of some 520,000 hectares.

The study encompassed four distinct agroecosystems:

- The head end of the left bank main canal command area of the TBP, an area supplied with abundant irrigation (*Head End*)
- The tail end of the command area, where irrigation is limited (*Tail End*)
- An area supplied with irrigation for the past 600 years through the ancient Vijayanagara canal system (*Ancient*)
- The rainfed area adjacent to the TBP (*Rainfed*)

Within each of these agroecosystems, three villages were chosen for detailed study. In each village, discussions were held with 10 farmers and their family members in order to gather information about farming practices, economic issues, domestic matters and a variety of environmental factors.

A central feature of the study was to obtain detailed quantitative energy, nutrient and economic information about all the activities and inputs (chemical products, biological materials, human labour, animal labour and mechanical devices) used in the agricultural production processes. Supporting information was acquired through government statistics, field and remote mapping, measurements of soil and water quality, studies of insect populations, a survey of pesticide use practices, a study of domestic activities, and discussions with older men.

Using the information obtained, the study examined sustainability under the six recommended headings of *productivity, stability, efficiency, durability, compatibility* and *equity*. A set of 35 quantitative indicators was developed, allowing for estimation and comparison of these categories of sustainability and of the overall sustainability within each system.

This approach enabled the identification of agricultural practices that are problematic in terms of their long-term sustainability. More positively, in other areas, highly productive yet environmentally and socially sustainable practices were also observed.

## Indicator data from the TBP area

Within each indicator category, individual values were scaled over a range of 0 to 10, with more sustainable measurements being assigned higher values. A summary of the findings for the four agroecosystems is given here.

- **Productivity:** Productivity is a measure of the average output of a given area, in terms of yield as well as in profit to the farmer. As has been emphasised, we maintain that a sustainable system must be a productive one in order to provide the farmer and his/her hired co-workers with an acceptable individual and family livelihood. A high level of productivity is also a prerequisite for feeding the wider population. Our indicators considered yield in the given area with respect to all-India benchmark values. They also measured crop value per cropping season per hectare and total farm income.

The agroecosystem values were as follows:

- Head End   8.5
- Tail End   3.5
- Ancient    8.7
- Rainfed    3.0

It is obvious and not surprising that productivity was enhanced in the areas well supplied with water. The high level of productivity is also a reflection of the knowledge and skill developed by the farmers in taking advantage of an irrigated agricultural situation.

- **Stability:** Stability measures the ability to maintain good levels of productivity over extended periods of time. To measure this historic yield, data for the past 20 years was used to determine variability in crop yields, a stable system being one that shows minimal variability. Soil properties in the various areas were also measured as a means of determining if the nutrient status was sufficient to support continued cropping. A common problem in irrigated areas is the development of waterlogging and salinity in poorly drained regions, and so a salinity index was included in the stability calculation.

- Head End   6.8
- Tail End   6.4
- Ancient    7.5
- Rainfed    5.7

Some level of stability was achieved by having an assured water supply through the irrigation systems. To some extent, however, this was offset by apparent nutrient depletion and encroaching soil salinity in the head end of the command area.

- **Efficiency:** Efficiency is the measure of the productive use of resources that make up the inputs used to grow a crop. Energy input/output ratios, renewable energy use and added nutrient (nitrogen) use were all determined. The productive areas again showed good energy use efficiency but most of the energy inputs were non-renewable forms. There was more efficient use of added nutrients and of renewable energy in the other settings.

  - Head End   3.9
  - Tail End   5.6
  - Ancient    5.8
  - Rainfed    5.8

- **Durability:** This category indicates a system's ability to resist environmental stress. In the TBP area the principal stressors are water (usually a lack of it) and insect pests. A durable system has an intrinsic ability to thrive even if the stress is present. Again, irrigation was shown to provide insurance against the possibility of lack of rainfall. However, in the tail end of the command area, the irrigation is itself somewhat uncertain. The situation with regard to insect pests is especially striking. In the head end, the average number of sprayings of paddy crops each year was found to be 18, while in the ancient area (with equivalent productivity) fewer than 2 sprayings were used. Intermediate values were measured in the other two agroecosystems. Information from the farmers in the Vijayanagara canal (Ancient) region indicated a widespread interest in more 'organic' methods of farming and a conscious decision to avoid overuse of non-renewable chemical inputs. Instead, they rely on methods of integrated pest management, especially by using varied cropping and crop rotation. The fact that in this area productivity was nearly equivalent to that in the more intensively farmed head end shows that productivity need not be compromised by those who consider environmental sustainability to be an important issue.

- Head End 5.1
- Tail End 6.3
- Ancient 9.1
- Rainfed 5.2

- **Compatibility:** In its broadest sense, compatibility is a measure of how well the agroecosystem fits in with its human, cultural and environmental surroundings. To measure it, two components of health were measured by evaluating the frequency of malaria and by determining potential human toxicity problems from the use of biocides. Ecological aspects of compatibility were assessed using remote sensing technology. This allowed the determination of the fraction of natural (uncultivated) area in each region, and the diversity of crops over time and space. Once again, there was a striking difference evident between the intensely cropped head end and the Vijayanagara canal areas in terms of the degree of biodiversity. This relates to the issues that relate to durability, as alluded to above. An unexpected finding was that the incidence of malaria was much lower in the irrigated areas than in the dryland. This may be due to the high concentrations of residual pesticides in the drainage canals.

- Head End 3.8
- Tail End 6.3
- Ancient 7.6
- Rainfed 6.0

- **Equity:** Equity was considered in a general sense as meaning the general well being of individuals, their families and the communities as a whole. An attempt was made to measure equity by indicators that relate to income ratios, education, health and house quality. The results of these assessments indicated generally low values, in large part as a result of the uniformly limited levels of education throughout the entire area. Once again, the ancient area stood out as having a better record than was observed in the other agroecosystems.

- Head End   4.8
- Tail End   3.6
- Ancient    5.1
- Rainfed    2.6

Taken together, the indicator set obtained in the study (Table 4.6) led to a number of useful conclusions.

Individual indicator values, as their name implies, are merely indicators of the sustainability of the agroecosystems in each area. As such, the actual values may be less important than what the values point to, and divergent numbers call for further, more detailed investigations. For example, the uniformly higher scores associated with the ancient area suggest that the systems of agriculture developed there over centuries make use of important principles of sustainability. This did not happen only by chance, as could be observed in discussions with a number of highly knowledgeable, articulate farmers. There was much to be learned from these people. On the other hand, the mediocre values in the head end region showed that while productivity at present is very high, there are signs that the system is ultimately unsustainable in its requirement for high levels of non-renewable chemical inputs and its dependence on the single,

Table 4.6
Sustainability matrix for the four agroecosystems of the Tungabhadra Project area. Summary (average) of indicator values within the six categories.

|    | Head | Tail | Ancient | Rainfed |
|----|------|------|---------|---------|
| Pr | 8.5  | 3.5  | 8.7     | 3.0     |
| St | 6.8  | 6.4  | 7.5     | 5.7     |
| Ef | 3.9  | 5.6  | 5.8     | 5.8     |
| Du | 5.1  | 6.3  | 9.1     | 5.2     |
| Co | 3.8  | 6.3  | 7.6     | 6.0     |
| Eq | 4.8  | 3.6  | 5.1     | 2.6     |

Source: From vanLoon et al. (2001b).

Pr – Productivity; St – Stability; Ef – Efficiency; Du – Durability; Co – Compatibility; Eq – Equity

**Poorest values**      **Best values**

presently-lucrative rice crop. The very poor index value for the rainfed region more than anything underlines the need for policies that support programs directed toward developing and implementing better methods for dryland farming.

Finally, it is important to be aware that the simple numbers obtained, while useful, cannot take the place of detailed qualitative studies in each of these areas. To some extent, it was possible to supplement the quantitative information by gathering other data through one-on-one discussions with a large number of persons living in each of the villages. The discussions gave substance to the numerical values. For example, in discussions with women throughout the command area, we found widespread concern for ensuring sustainability, especially in terms of family life. Their domestic activities too play an important role in overall village life. Even in terms of energy consumption, cooking activities are one of the main consumers of energy, and this leads to its own set of questions about how a particular domestic practice relates to sustainability. Discussions with both women and men evoked concerns about health issues, connected with chemical use and with changing lifestyles that evolve from new methods of agriculture.

Therefore, while we promote the use of numerical indicators, we also emphasise that these should be used only as a simple means by which it becomes possible to focus on issues which are generally very complex and not described in simple categories.

## Policy

The TBP study leads us to say something about policy implications.

- We encourage the development and use of micro-indicators that are appropriate to the Indian agricultural situation. This and other similar studies should be considered as initial steps directed toward providing a set of proposed indicators. In the TBP assessment, indicators were formulated based on the concept that a healthy rural environment is essential for India in the twenty-first century. Information that could be collected by a small team over a period of several months was proposed, but refinement and modification of the protocol will continue to be required.

- Information such as was obtained in the study should be made available to government departments and non-governmental organisations that are dealing with agriculture and the environment. In some cases, these departments are measuring macro-sustainability and the shared micro information can assist in adding detail to the broader picture. At the same time, the regional and national information points to issues that may have been neglected in the local assessments.
- Based on this present study, it is recommended that indicators be one factor to be used in determining agricultural policy in the regions of India. For example, the following proposals for India are a natural outcome emanating from a simple interpretation of the data:
    o Major efforts should be made to reduce dependence on pesticides in situations of high-intensity agriculture.
    o State governments should establish training and regulatory systems that will lead to the safe use of pesticides.
    o Management of irrigation water resources should be improved to provide a more equitable distribution system, as well as one that encourages cropping diversity. Much more effort should be directed toward enhancing sustainable agriculture technologies for rainfed areas.
    o Steps should be taken to support the use of wealth generated by agriculture for community (as opposed to individual) services, such as provision of better education and health facilities, higher quality water supply, and cleaner, more sanitary living environments.
    o Low non-renewable input farming methods such as those used in the Vijayanagara area should be promoted as a good example of how sustainable, high-productivity agriculture is possible in the Indian setting.

## 4.8 Assessments of Sustainability over Broad Regions

The emphasis in this book has been to develop ways of assessing agricultural operations at the local or micro level. Assessment, however, is

## Studies of Sustainability

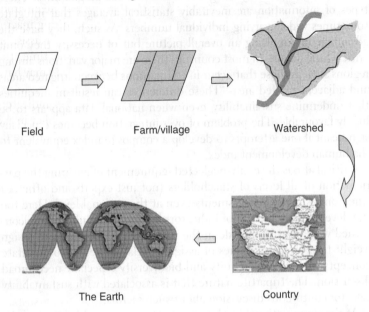

**Figure 4.2**
Levels at which agricultural sustainability can be measured

required at higher levels of geographic aggregation—from farm onwards, through village, project area, ecologically defined regions like watersheds, to regions that are politically defined, such as states or nations (see Figure 4.2). The more local information provides details about regional forces and activities that shape management decisions on the ground, while the macro information may reflect the broader features of climate, commodity prices and national and global politics. Understanding the relations and interactions between the diverse factors is a particular challenge at all levels of assessment—a challenge that is essential to meet if one is to make rational decisions that take into account both local realities and global forces.

In certain ways, the development of indicators at a macro-level is quite different and in some ways easier from the process when it is done locally. For example, in the case of national indicators, it is usually possible to make use of scientific and statistical data that are readily available. This includes information about productivity, water supply, total use of various manufactured resources like fuels, fertilisers and other chemicals, and data regarding incomes and social programmes. These types of information are inevitably statistical averages that integrate sometimes widely varying individual numbers. As such, they have the advantage of supplying an overall picture but of necessity they omit critical local details. In most countries there are major variations among regions such as those that occur in comparisons between irrigated areas and adjacent rainfed areas. These differences can result in inequities that undermine sustainability, even when national data appears to be highly favourable. The problem of over-integration becomes especially significant if one attempts to develop a composite index equivalent to the human development index.

Critical also is the easily-neglected requirement of ensuring the participation of all levels of stakeholders (not just experts and officials) when one is making assessments even at the national level. Here too the development and use of indicators involve value judgements concerned with what to include in the measurements and how to assign weight to each factor. Issues of assigning relative weight to disparate concepts such as social equity and biodiversity especially need broad discussion. The tripartite nature that is associated with sustainability calls for continuous discussion about such factors.

At the supra-national level, new sets of issues are added to the mix. Especially in this regard, decisions about equity in access to opportunities and resources become critical. Table 4.7 summarises issues associated with assessment of agricultural sustainability at various levels.

Using our set of six categories, macro-level sustainability indicators can be developed and would take the following general form:

## Productivity

Indicators used at the micro-level can in many cases be modified for use on a larger scale. Average values of yield for individual crops allow for comparisons made over time and with other countries. A composite index can be constructed by weighting yields by the fraction of land devoted to each crop giving, for example, an average grain yield within a particular country. Another indicator related to food availability and security is the total production per capita of staple crops produced in the country.

Table 4.7
Levels of agricultural sustainability assessment

| Level of assessment | Characteristics of agricultural sustainability | Assessment categories |
|---|---|---|
| Field | Productive crops, soil/water health, limited non-renewable inputs, crop diversity, limited pest problems | Productivity, stability, efficiency durability, compatibility |
| Village | As above, but also support for wild biodiversity, community social facilities | Productivity, stability, efficiency durability, compatibility, equity |
| Country | Public awareness and support for agroecology and conservation of resources | Agroecological policies and implementation, pollution due to agriculture, resource use rural/urban equity |
| Continent/Earth | Quality natural environment, stable climate, food security, equitable trading system | Pollution associated with agriculture, sustainable population and resource supplies, global equity |

Likewise, it is possible to measure average incomes of farmers within a country.

## Stability

Reliable data are available regarding some forms of land degradation such as desertification and stability. An index based on a time series showing percentage of land lost to such forms of degradation provides useful information (in a negative sense), regarding long-term prospects for maintaining a stable food production system. Less easily obtained are data regarding some other forms of land degradation such as losses due to erosion. Losses of productive land that result from expansion of urban, industrial and other infrastructure are equally important in terms of food security.

Stability of water supply can be measured by a time series of rainfall patterns within the country. Regional information regarding major groundwater reserves is also commonly available. Annual water withdrawal compared with water available in a major region or nation is a good indicator of sustainability of the resource.

As with water, energy security is an essential subject for any country, and annual consumption compared to availability is a measure of the long-term prospects in terms of a stable supply.

## Efficiency

On a national scale, it is possible to measure total fertiliser (or specific nutrient) use per hectare. With these measures, nutrient use efficiency can be calculated from the ratio of nutrient removed in the harvested crops (weighted by fraction of crop) divided by the total nutrient minus fertiliser source of nutrient. Detailed energy data that can be used over broad regions may not be readily available. Efficiency in terms of crop output per farm worker can be calculated using demographic and production information.

Water use efficiency indices can also be calculated, especially in rainfed areas. In such cases, there will be detailed regional information regarding volumes of water available to farmers. Efficiency in terms of comparative productivity on irrigated as well as non-irrigated lands can also be calculated.

## Durability

Data on pesticide use can be obtained from production and sales statistics; this information is valuable in itself and can be used along with cropping information to calculate an index of average pesticide use per hectare.

Additional indicators related to resistance of pests to pesticides can be evaluated, e.g., the number of known instances of pest resistance weighted by the portion of land affected by the pest resistance.

## Compatibility

Many of the indices of diversity described earlier, whether of agricultural crops, of wild species or of landscape types, can be readily measured. Satellite images of various types can provide information at a surprising level of detail. In fact, diversity measurements based on data obtained over wider regions are expected to be highly reliable and meaningful, since their statistical validity improves with the size of the sample.

As more information becomes available regarding the contribution of various land use activities to greenhouse gas production, it is becoming increasingly possible to accurately measure national releases of greenhouse gases associated with agriculture. The data associated with agronomy can then be supplemented with information from other agricultural practices such as livestock production. The data obtained provide information that relates to compatibility with the overall global environment.

Many response indicators can be developed related to initiatives that are designed to preserve natural space, to protect endangered species, and to promote sustainable methods of agriculture.

## Equity

The United Nations Human Development Report (1992) clearly documents that during the 1970s and 1980s, a period of unprecedented economic growth on a global scale, the gap between rich and poor had widened. The subsequent decade into the twenty-first century has seen this trend continuing at an accelerating rate. Inequalities in the rural sector are especially evident.

Education and health statistics are collected in almost every country. There may however be difficulties in factoring out the information that relates to rural communities, since data usually relates to a region that includes both rural and urban areas. Income data are frequently categorised by profession, so that a distinction can be made between agricultural and other income figures.

The average income data can be supplemented by measures of distribution, using categories similar to those recommended for use at the micro-scale. Equity in income for agricultural labour compared with other forms of labour can be determined.

Nutritional availability at the national level is calculated as the daily energy supply available (from food grains) per person, often expressed as a percentage of energy requirements. Likewise, protein availability per capita can be estimated, based on quantitative information regarding components of the national food basket.

In some countries, there is also solid information regarding gender differentials in education, health (often measured by life span), although once again this may not be disaggregated into rural and urban statistics.

# Bibliography

Aggarwal, G.C. 1989. 'Judicious Use of Dung in the Third World', *Energy* 14: 349–52.
Allen, R.G., L.S. Pereira, D. Raes, and M. Smith. 1998. *Revised FAO Methodology for Crop Water Requirements in Irrigation and Drainage, Paper no. 56: Crop. Evapotranspiration.* Rome: Food and Agriculture Organization of the United Nations.
Altieri, M. 2001. 'Account of the Smallholder Workshop', *Ecology and Farming* 27.
Altieri, M.A. 1987. *Agroecology: The Scientific Basis of Alternative Agriculture.* Boulder: Westview Press.
Anonymous. 2000. *World Resources 2000–2001: People and Ecosystems.* Washington D.C.: World Resources Institute.
Campbell, B.M., P. Bradley and S.E. Carter. 1997. 'Sustainability and Peasant Farming Systems: Observations from Zimbabwe', *Agriculture and Human Values* 14: 159–68.
Chesworth, Ward, Michael R. Moss and Vernon G. Thomas (eds.). 2002. *Sustainable Development: Mandate or Mantra.* Faculty of Environmental Sciences, University of Guelph.
Club of Rome. 1972. *The Limits to Growth.* London: Earth Island Limited.
Conway, G. 1997. *The Doubly Green Revolution.* London: Penguin Books.
Dahiya, A.K. and P. Vasudevan. 1986. 'A Field Study of Energy Consumption Patterns on Small Farms', *Energy* 11: 685–89.
Daly, H.E. and J.B. Cobb. 1990. *For the Common Good.* London: Green Print.
Daly, H.E. 1997. *Steady-state Economics.* San Francisco: Freeman.
Damodaran, A. 2001. *Toward an Agro-ecosystem Policy for India.* New Delhi: Tata McGraw-Hill Publishing Co. Ltd.
Doran, J.W., D.C. Coleman, D.F. Bezdicek and B.A. Stewart. 1994. *Defining Soil Quality for A Sustainable Environment.* Soil Science Society of America Special Publication Number 35, Soil Science Society of America, Inc.
Dresner, Simon. 2002. *Principles of Sustainability.* London: Earthscan Publications Ltd.
Edwards, S. 2002. 'A Project on Sustainable Development through Ecological Land Management by Some Rural Communities in Tigray', Tigray Institute for Sustainable Development, Unpublished Document.
ESCAP (Economic and Social Commission for Asia and the Pacific). 2002. *Sustainable Social Development in a Period of Rapid Globalization: Challenges, Opportunities and Policy Options.* New York: United Nations.
Fluck, R.C. and C.D. Baird. 1980. *Agricultural Energetics.* Westport: AVI Publishing.
Gliessmann, S.R. 1998. *Agroecology: Ecological Processes in Sustainable Agriculture.* Chelsea: Ann Arbor Press.

Gupta, R.S.R., H.S. Malik, R.K. Malik and A.R. Rao. 1984. 'Energy Utilization in Maize Production', *Energy* 9: 189–92.

Gupta, R.K., N.T. Singh and M. Sethi. Undated. 'Ground Water Quality For Irrigation in India', Technical Bulletin No.19, Central Soil Salinity Research Institute, Karnal, India.

Gustafson, R.L. 1989. 'Groundwater Ubiquity Score: A Simple Method for Assessing Pesticide Leachability', *Environmental Toxicology and Chemistry* 8: 339–57.

Hall, C.A.S. 2000. *Quantifying Sustainable Development: The Future of Tropical Economics*. San Diego: Academic Press.

Harris, R.F. and D.F. Bezdicek. 1994. 'Descriptive Aspects of Soil Quality/Health', in Doran, J.W., D.C. Coleman, D.F. Bezdicek and B.A. Stewart, *Defining Soil Quality for A Sustainable Environment*, Soil Science Society of America Special Publication Number 35, Soil Science Society of America, Inc.

Heinonen, E. 2001. *Sustainability in Agriculture: How to Define it and Can it be Measured?* At http://www.oac.uoguelph.ca/FSR/collection/indicator/indexproposal/ 0.5txt.

IPCC (Intergovernmental Panel on Climate Change). 2001. *Climate Change 2001: The Scientific Basis*. Cambridge University Press.

Jansen, D.M., J.J. Stoorvogel, and R.A. Schipper. 1993. 'Using Sustainability Indicators in Agricultural Land Use Analysis: An Example from Costa Rica', *Netherlands Journal of Agricultural Science* 43: 61–82.

Jones, H. 1965. *John Muir and the Sierra Club: The Battle for the Yosemite*. San Francisco: Sierra Club Books.

Kettel, B. 2001. *Gender Sensitive Indicators: A Key Tool for Gender Mainstreaming*. SD Dimensions, Sustainable Development (SD) Department, FAO.

King, Franklin Hiram. 1927. *Farmers of Forty Centuries: Permanent Agriculture in China, Korea and Japan*. London: Jonathan Cape Limited. First published in 1911.

Landes, David S. 1999. *The Wealth and Poverty of Nations*. New York: W.W. Norton & Company, Inc.

Lefroy, R.D.B., H.D. Bechstedt and M. Rais. 1999. 'Indicators for Sustainable Land Management based on Farmer Surveys in Vietnam, Indonesia and Thailand', *Agriculture, Ecosystems and the Environment*.

Leigh, R.A. and A.E. Johnston. 1994. *Long-term Experiments in Agriculture and Ecological Sciences*. Oxford: Oxford University Press.

Loomis, R.S. and D.J. Connor. 1992. *Crop Ecology: Productivity and Management in Agricultural Systems*. New York: Cambridge University Press.

Makhijani, A.B. and A.M. Lichtenberg. 1972. 'Energy and Well-being', *Environment* 14: 10–18.

Meadows, D. 1998. *Indicators and Information for Sustainable Development*. Hartland Four Corners, Vermont: The Sustainability Institute.

Meerman, F., G.W.J. Van de Ven, H. Van Keulen and H. Breman. 1996. 'Integrated Crop Management: An Approach to Sustainable Agricultural Development. *International Journal of Pest Management* 42: 13–24.

Mellon, M., J. Rissler and F. McCamant. 1995. Union of Concerned Scientists Briefing Paper.

Molden, D., R. Sakthivadivel, C.J. Perry, C. de Fraiture and W.H. Kloezen. 1998. 'Indicators for Comparing Performance of Irrigated Agricultural Systems', Research Report 20, International Water Management Institute, Colombo, Sri Lanka.

O'Brien, Paul. 2001. *Encouraging Environmentally Sustainable Growth in the United States*. Organization for Economic Cooperation and Development (OECD), Economics Department Working Paper No. 278, p. 22.

# Bibliography 275

Parrott N. and T. Marsden. 2002. *The Real Green Revolution: Organic and Agroecological Farming in the South*. Canonbury Villas, London: Greenpeace Environmental Trust.

Pimentel D. (ed). 1980. *CRC Handbook of Energy Utilization in Agriculture*. Boca Raton: CRC Press.

Pimentel, D. and M. Pimentel. 1979. *Food, Energy and Society*. London: Edward Arnold.

Ponting, Clive. 1991. *A Green History of the World*. New York: Penguin Books.

Prasad, J. 1995. *Goat, Sheep and Pig Production and Management*. Ludhiana: Kalyani Publishers.

Pretty, J. 1999. 'Can Sustainable Agriculture Feed Africa? New Evidence on Progress, Processes and Impacts', *Environment, Development and Sustainability* 1: 253–74.

Qizilbash, M., 2001. 'Sustainable Development: Concepts and Rankings', *Journal of Development Studies* 37: 134–61.

Rao, A.R. 1985. 'Fertilizer Use Saves Fuels in India' *Energy* 10: 989–91.

Rao, A.R., K.S. Nehra, D.P. Singh and M.S. Kairon. 1992. 'Energetics of Cotton Agronomy', *Energy* 17: 493–97.

Rogers, P.P., K.F. Jalal, B.N. Lohani, G.M. Owens, C-C Yu, C.M. Dufournaud, and J. Bi. 1997. *Measuring Environmental Quality in Asia*. The Division of Engineering and Applied Sciences, Harvard University and the Asian Development Bank.

Romig, D.E., M.J. Garlynd, R.F. Harris and K. McSweeney. 1995. 'How Farmers Assess Soil Health and Quality', *Journal of Soil and Water Conservation* 50: 229–36.

Sachs, Wolfgang, Loske Reinhard and Manfred Linz. 1998. *Greening the North*. London: Zed Books.

Segnestam, L. 2000. *Developing indicators: Lessons Learned from Central America*. Washington D.C.: The International Bank for Reconstruction and Development/The World Bank.

Shyam, M. and L.P. Gite. 1990. 'Energy Analysis of Soyabean–Wheat Crop Rotation under Rainfed Conditions of Central India', *Energy* 15: 907–12.

Smeets, E. and R. Weterings. 1999. 'Environmental Indicators: Typology and Overview', Technical Report 25, European Environment Agency (EEA), Copenhagen.

Smil, V. 1993. *Biomass Energies*. New York: Plenum Press.

Stocking, M.A. and N. Murnaghan. 2001. *Handbook for Field Assessment of Land Degradation*. London: Earthscan Publications Ltd.

Taylor, D.C., Z.A. Mohamed, M.N. Shamsudin, M.G. Mohayidin and E.F.C. Chiew. 1993. 'Creating a Farmer Sustainability Index: A Malaysian Case Study', *American Journal of Alternative Agriculture* 8: 175–84.

Thompson, A.B., G.W. vanLoon, L.B. Hugar and S.G. Patil. 2001. 'Application of Remote Sensing Technology for Assessing Crop Diversity in Four Agricultural Systems of North Karnataka', in Muralikrishna T.V. (ed.), *Spatial Information Technology—Remote Sensing and Geographical Information Systems, Volume II*. B.S. Publications, p. 288.

United Nations Development Programme. 1992. *Human Development Report, 1992*. Oxford: Oxford University Press.

USSL (United States Salinity Laboratory). 1954. *Diagnosis and Improvement of Saline and Alkaline Soils*. USDA Handbook No. 60. Washington, D.C.: USSL

vanLoon, G.W., S.G. Patil and L.B. Hugar. 2001a. 'Comparative Methods for Assessing Real Costs and Benefits of Different Agricultural Systems in Selected Villages in South India', Final Report to the Shastri Indo-Canadian Institute, New Delhi.

vanLoon, G.W., S.G. Patil, L.B. Hugar, M.S. Veerapur, J. Yerriswamy, T. Cross and A.C. vanLoon. 2001b. 'Assessment of Agricultural Sustainability in Four Agroecosystems in the Tungabhadra Project Area of South India'. Presented at International Conference on Managing Natural Resources for Sustainable Agricultural Production in the 21st Century, New Delhi. Volume 3: Resource Management, 1395–97.

Verma, S.R. and B.S. Pathak. 1998. *Tractor Maintenance and Operation*. Ludhiana: Communication Centre, Punjab Agricultural University.

Visser, S. and D. Parkinson. 1992. 'Soil Biological Criteria as Indicators of Soil Quality: Soil Microorganisms', *American Journal of Alternative Agriculture* 7: 31–37.

Wackernagel, M. and W. Rees. 1996. *Our Ecological Footprint*. Gabriola Island, BC: New Society Publishers.

Wackernagel, M., Larry Onisto, Alejandro Callejas Linares, Ina Susana López Falfán, Jesus Méndez García, Ana Isabel Suárez Guerrero and Ma. Guadalupe Suárez Guerrero. 2002. 'Ecological Footprints of Nations: How Much Nature Do They Use? — How Much Nature Do They Have?' At http://www.ecouncil.ac.cr/rio/focus/report/english/footprint/.

WCED (World Commission on Environment and Development). 1987. *Our Common Future*. Oxford: Oxford University Press.

White, L. Jr. 1967. 'The Historical Roots of Our Ecologic Crisis', *Science* 155: 1203.

WHO (World Health Organization). 2003. *Guidelines for Drinking Water Quality, 3rd Edition*. Geneva: WHO.

# Index

Agricultural capital, forms of, 47–49
Agricultural ecosystems (agroecosytems), 52–54
Agricultural sustainability, 34–55; assessing, 44–49; categories, 64–65
Agriculture, conventional, 123–124; low input, 123–124; agricultural process, 155

Benchmark, 93–95, 111
Biodiversity, 203–211
Biomass, 49–50, 107, 157; animal manure, 173; as animal fodder, 174; as fuel, 176–181, 228–229; as soil amendment, 174–176, 211–212; conversion, 178–181; microbial, 143; secondary, 171–185; yield, 113–114
Biogas, 180–181

Capital, forms of, 48, 153–154
Carbon dioxide emissions, 85–86
Cash crops, 118–119
Central America, 255–258
China, 46
Club of Rome, 73–74
Compatibility, 65; assessment strategy, 212; cultural, 202–203; indicators, 196–212; in South-East Asia, 244–247; in the Tungabhadra area, 260; over broad regions, 270
Composite index, see Index
Conceptual overview, 63–65
Connectedness, of natural areas, 210
Conservationalists, 23–24

Development, 27
Durability, 65; assessment strategy, 195–196; economic stress, 194–195; indicators, 185–196; in South-East Asia, 244–247; in farming systems in Malaysia, 242–243; in the Tungabhadra area, 261–262; over broad regions, 270; pest stress, 189–194; water stress, 187–189

Earthworms, 143
Ecological footprint, 75–76
Economy, 31–32, 40
Ecosystems, 49–55, 196–197
Education, 99–100, 217–219
Efficiency, agricultural, 64; assessment strategy, 185; energy as a measure of, 159–169; indicators, 153–185; in South-East Asia, 244–247; in the Tungabhadra area, 260; nutrient use, 161–171; over broad regions, 270; water use, 158–159, 251–255
Energy, 36, 50–53, 74; efficiency, 159–169; nutrient use, 167–171; table of equivalents, 161–164

Environment, 30–31, 39
Equity, 65; assessment strategy, 253; education, 217–219; food and nutrition, 229–232; gender, 224–226; health, 226–229; indicators, 213–253; income, 219–224; in South-East Asia, 244–247; in the Tungabhadra area, 260–263; over broad regions, 270
Erosion, see Soil
Europe, 249–250

Food security, 20, 267
Food and nutrition, 35–36, 229–232

Gender, 69, 224–226
Genetically modified organisms (GMOs), 190
Goalposts, 89–91
Global assessments of sustainability, 265–271
Global food production, 35–38
Green revolution, 19

Human development index, 98–102
Hunting and gathering, 21–22
Health, 99, 226–229

Income, 119–122, 219–224
Index, combining indicator values, 95–102, 234–236
India, 259–265
Indicators, availability and collection of data for, 66–67; categories, 64–65; composite, 95–98; diagrams, 103–104; integrative capability, 76–78; predictive ability, 80–82; productivity, 106–124; properties, 65–84; purpose, 57–60; quantification, 72–76, 84–104; relevance, quality and reliability, 65–66; secondary, 62; sensitivity to changes, 78–80; stability, 124–153; stakeholder participation, 68–72; steps in developing, 63–65; types, 60–63
Indonesia, 243–247
Intercropping, see multicropping
Irrigation, 147–151, 188–189, 206–209, 214, 259–265

Malaysia, 242–243
Microorganisms, 49–50; in soil, 142–143
Monoculture, 114–115
Mixed cropping, see multicropping
Multicropping, evaluation of productivity, 114–118, 204–205

Nutrition, 36, 114

Organic matter, 157, 239

Participatory methods, 71–72
Pesticides, 189–192
Photosynthesis, 49, 157
Policy, 265–266
Preservationists, 23
Productivity, 64, 106–124; assessment strategy, 122–123; indicators, 106–124; in irrigated systems, 251–255; in South-East Asia, 244–246; in the Tungabhadra area, 261; monetary indicators, 119–122; potential, 108–111; over broad regions, 267–268
Protein, 114–115
PSR system, 60–63

Rainfall, 146–147, 148, 187–189
Rothamsted agricultural research station, 125–126

Society, 32, 40–41
Soil, biological properties, 142–144; chemical properties, 140–142; degradation, 107, 127–128; erosion, 128–131, 239; organic matter, 140, 240, 244; physical

properties, 139–142; quality, 131–144, 243; quantity, 128–131
Sri Lanka, 251–255
Stability, 64; assessment strategy, 152; indicators, 124–153; in farming systems in Zimbabwe, 239–241; in Malaysia, 242; in South-East Asia, 244–245; in the Tungabhadra area, 261–262; over broad regions, 269–270
Stakeholders, 68
Sustainability, 28; agricultural sustainability, 34-55, 44; levels of, 41–43; sustainability tripod, 30–34
Sustainable development, 19–34; background and history, 21–23

Thailand, 244–248
Tungabhadra project area, 206–209, 222–223, 259–266

Vietnam, 244–248

Water, consumption, 94, 145; degradation, 127–128; for domestic consumption, 199–202; groundwater, 127–128, 147–148; irrigation, 148–152; quality, 149–152, 199–202; quantity, 146–148; stress, 187–189

Yield, 109–119; variability of, 127

Zimbabwe, 239–242

# About the Authors

**Gary W. vanLoon** is Professor in the Department of Chemistry, and Associate Director of the School of Environmental Studies, Queen's University, Kingston, Ontario, Canada. A member of the Department since 1969, Dr vanLoon's research mainly focuses on environmental chemistry, particularly issues related to soil-water relations. He has published numerous articles in various journals, and is also the co-author of *Environmental Chemistry—A Global Perspective* (with Stephen J. Duffy, 2000). A founding member of the School of Environmental Studies at Queen's University, Dr vanLoon has over the years been involved in research and environmental projects in India, Nigeria and South Africa.

**S.G. Patil** is Professor and Head, Department of Environmental Sciences, University of Agricultural Sciences, Dharwad, India. He is also Chief Scientist on the All India Coordinated Project on Water Management. Dr Patil has contributed a large number of articles to various journals, and is one of the editors of a book on 25 years of research in the 'Management of salt-affected soils and use of saline water in agriculture' published by the Indian Council of Agricultural Research. His main areas of interest are salinity management, identification of salt tolerance in crop plants and forest/horticultural species, and bio-drainage.

**L.B. Hugar** is Professor in the Department of Agricultural Economics, College of Agriculture, Raichur, India. A member of the Department for over 24 years, Dr Hugar's areas of interest are natural resource management, farm management economics and agricultural marketing.

His work on canal-irrigation water management issues in the Upper Krishna Project formed the basis for the amendment of the Irrigation Act of 2003 by the Government of Karnataka. Dr Hugar has published numerous articles in various journals, and presented over a dozen research papers at national and international seminars and symposia.